Making the Most of the Postdoc

Graduate students and postdoctoral fellows spend upwards of 15 years honing their research skills. However, in all this training, compulsory career and professional development courses are far and few between. In the absence of a formal training curriculum, this co-curricular postdoc guidebook can be used as a manual for aspiring scientists to find career success.

Postdocs face many hurdles in their pursuit of research excellence and independence. None more frustrating than making the most of this challenging yet rewarding opportunity. Ultimately, the point of postdoc training is not maintaining a lengthy postdoc tenure but landing a satisfying job. Regardless of what they do in their career, postdocs need to gain and master many skills both directly related to their scientific training and beyond. This book posits that if trainees are motivated and given some practical guidance, they can build a professional reputation while achieving a successful postdoc experience.

Based on the personal experiences of the author, this book logically outlines the flow of the postdoc experience from beginning to end by providing actionable advice on how to get the most out of postdoctoral training while laying out strategies for choosing the right research environment to thrive along with planning, and executing, a successful postdoc tenure. Written for current and future postdocs, as well as their mentors, this book covers what they need to know, and do, to strategically advance in their early research career.

Key Features:

- Practical and actionable advice from an author that has experienced PhD and postdoc training and is now directing a postdoc office at a world-renowned research institution
- Methodical approach most readers can readily adapt for their own purposes
- Specifically written for current and future STEM postdocs while being agnostic of the research field

Dr. James Gould, PhD is Director of the HMS/HSDM Office for Postdoctoral Fellows at Harvard Medical School (HMS) where he has implemented research, career, and professional development programs and policies for HMS-affiliated trainees since 2011. Prior to HMS, Dr. Gould completed two postdoc fellowships at the National Cancer Institute of the NIH where he became involved in training affairs and studied cancer metabolism. Dr. Gould received his BS in Biotechnology/Molecular Biology from Clarion University of Pennsylvania and his PhD in Biochemistry and Molecular Biology from the University of Louisville.

Making the Most of the Postdoc

Strategically Advancing Your Early Career

James Gould

CRC Press
Taylor & Francis Group
Boca Raton London New York

CRC Press is an imprint of the
Taylor & Francis Group, an **informa** business

Designed cover image: © Shutterstock, ID 786339889, Vector Contributor Artistdesign29

First edition published 2023
by CRC Press
6000 Broken Sound Parkway NW, Suite 300, Boca Raton, FL 33487-2742

and by CRC Press
4 Park Square, Milton Park, Abingdon, Oxon, OX14 4RN

CRC Press is an imprint of Taylor & Francis Group, LLC

© 2024 James Gould

Reasonable efforts have been made to publish reliable data and information, but the author and publisher cannot assume responsibility for the validity of all materials or the consequences of their use. The authors and publishers have attempted to trace the copyright holders of all material reproduced in this publication and apologize to copyright holders if permission to publish in this form has not been obtained. If any copyright material has not been acknowledged please write and let us know so we may rectify in any future reprint.

Except as permitted under U.S. Copyright Law, no part of this book may be reprinted, reproduced, transmitted, or utilized in any form by any electronic, mechanical, or other means, now known or hereafter invented, including photocopying, microfilming, and recording, or in any information storage or retrieval system, without written permission from the publishers.

For permission to photocopy or use material electronically from this work, access www.copyright.com or contact the Copyright Clearance Center, Inc. (CCC), 222 Rosewood Drive, Danvers, MA 01923, 978-750-8400. For works that are not available on CCC please contact mpkbookspermissions@tandf.co.uk

Trademark notice: Product or corporate names may be trademarks or registered trademarks and are used only for identification and explanation without intent to infringe.

ISBN: 978-1-032-24678-9 (hbk)
ISBN: 978-1-032-25886-7 (pbk)
ISBN: 978-1-003-28545-8 (ebk)

DOI: 10.1201/9781003285458

Typeset in Palatino
by Deanta Global Publishing Services, Chennai, India

Contents

Foreword vii
Acknowledgments ix

Part 1 The Beginning

Introduction: My Path – from Clueless to Clarity: Finding Strength through Struggle 3

1 How to Use This Book 11
    Are You Puzzled? 11
    How to Use This Book If You Are a Postdoc 12
    How to Use This Book If You Are a Faculty Mentor 13
    How to Use This Book If You Are a Postdoc Affairs or Career Office Leader 14
    How to Use This Book to Improve the Research Enterprise 15

2 The Most Important Professional Decision You Will Make … So Far 17
    To Postdoc or Not to Postdoc 17
    Postdoc Job Search 19
    Creating a Productive Environment 23

3 The Postdoc Protocol 25
    Postdoc Process 25
    Envisioning the Endpoint 25
    Self-Assessment and Reflection 27
    Developing Plans and Accomplishing Goals 29
    Maintaining Progress 32
    Getting a Job 33

Part 2 The Middle

4 Situational Awareness 37
    Postdoc Pain/Pivot Points 37
    Stopping the Negativity Spiral 41
    Imposter Syndrome: Managing Your Inner Dialogue 43
    Resilience and Mental Wellness 44

5 Navigating through Your Postdoc 47
Establishing Ground Rules and Expectations 47
Mentorship and Individual Development Plans (IDPs) 49
Mentoring Up and Self-Advocacy 51
Professional Development 52

6 Path to Independence 55
Understanding Your Priorities 55
Framing Your Training 55
Your Training Efforts 56
Building Your Reputation 57
Build Transferable Skills 59
Preparing for the Next Step 60
Structuring Your Preparation 61

Part 3   The End

7 Career Transition Readiness 65
Knowing When You Are Ready to Leave 65
Navigating an Unknown Process 65
Activating Your Network 67
Common Ground 67
Contact Points 68
Integrating Networking, Research, and Life 70
Having "The Talk" with Your Mentor 71
Launching Your Job Search 72

8 Some Advice on Application Materials 75
Applications and Job Postings 75
Cover Letters 77
Crafting the Résumé and CV 79
Statements of Research and Teaching 88

9 Interview Preparation 91
Becoming a Storyteller 91
Common Struggles and Successful Strategies for Interview Prep 95
Interview Performance 96
Do Your Homework 97
The Face-to-Face Interview 98
Gratitude and Follow-Up 99

10 Negotiating Your Exit 101
Considering and Negotiating the Offer 101
Continuing "The Talk" with Your Mentor 103

Index 107

Foreword

Getting a PhD is a special journey. One that should unite us all together as PhDs. Unfortunately, the nature of our work tends to fragment us at times. We can sometimes be critical of ourselves and other PhDs. Understand that it was never meant to be this way. We were meant to be critical of our research, but not our PhD peers, mentors, or trainees. We were meant to be advocates of other PhDs.

Dr. James Gould is an amazing example of somebody who has dedicated his life to being an advocate for other PhDs, specifically those in postdoctoral appointments. Dr. Gould, who I will call my great friend Jim from here on out, is not just an advocate for the PhDs he mentors as Director of the Harvard Medical School and Harvard School of Dental Medicine Office for Postdoctoral Fellows, a very prestigious role he earned through hard work, intelligence, kindness, determination, and an incredible ability to teach and guide others; he is also an advocate for his PhD peers.

When I was first starting out in industry after getting my PhD, I was living in Boston and decided to start my company, Cheeky Scientist. I was eager to speak to PhDs at local universities and help PhDs learn how to transition out of academia and into industry careers. This was the company's mission. The problem? No one was interested. Until Jim. Jim gave me a chance to speak in my very first seminar under the Cheeky Scientist banner. After a mountain of rejections, he said yes and I couldn't believe it. This was ten years ago in 2013 and the idea that PhDs should be trained to do something other than become professors was not widely accepted. There is still a lot of work to be done in terms of its acceptance today, but a decade ago, whenever I spoke to a room of 50 or so PhDs and asked who among them was going to stay in academia, 40 would raise their hands. Today, that ratio is flipped, with 40 raising their hands to indicate they are pursuing careers outside of academia.

Jim was one of the few people to see this coming. He was also the first person to see potential in me as a fellow mentor of PhDs instead of a threat. When Jim invited me to speak, or should I say, when he allowed me to come speak at Harvard to the postdocs he was mentoring, he did so without judgment, reservations, caveats, or complaints. Instead, he invited me with friendliness, open arms, and a collegial spirit. He complimented me on the message I was trying to get across and gently guided me on making my presentation less abrasive and more considerate and inclusive to academics, many of whom were skeptical of anyone suggesting they look for work outside of the academy.

Jim continued to offer his advice to me over the years, always gently, intelligently, and without ego, helping me understand the challenges postdocs faced and how to gain wider acceptance of my ideas by smoothing out the rough edges of messages. Without Jim, the message, voice, and even the brand of Cheeky Scientist wouldn't be where it is today. After speaking at hundreds of universities, all around the world, there are only a few people I have worked with who get as much done for PhDs as Jim, and there is no one who genuinely cares about postdocs as much as him. Jim has deep knowledge of the terrain in terms of what a postdoc goes through psychologically, the pressures they must deal with, how many self-sabotage their careers for years, and, most importantly, the world of possibilities open to postdocs, career-wise and otherwise, once they expand their vision of their training, their skills, and their worth.

Boston is the beating heart of PhDs worldwide, with more PhDs in Boston and Cambridge than any other metropolitan area in the US, and possibly the world. Jim is at the center of that beating heart. If you are doing a postdoc, or are going to do a postdoc, this book is critical to you making the most of your time in your position and then leveraging your training to get hired into a meaningful career. Jim will show you in Chapter 4 what your professional pain points are or will be as a postdoc and what professional pivot points will alleviate those pain points. In Chapter 9 he will show you how to use the P-A-R Stories and the P-A-R Matrix to interview for your next role successfully. From thriving as a postdoc to arriving in the next step in your career, *Making the Most of the Postdoc* will show you the way.

I want to encourage you to read Jim's words with a discovery mindset. Open yourself up to possibilities. Start this book as an exploration. After all, you likely got into your PhD in the first place to discover and to explore. Savor the adventure that you're about to go on. And then commit to putting Jim's words into action for yourself and your career. Enjoy!

Dr. Isaiah Hankel
Founder and CEO of Cheeky Scientist
Author of *The Power of a PhD*, *Intelligent Achievement*, and *Black Hole Focus*

Acknowledgments

Thank you to everyone who helped me in thinking about, organizing, drafting, editing, and submitting this book. It was a true community effort full of support and enthusiasm. I am eternally grateful to my wife, Heidi, and daughter, Lucy, for their steadfast belief in my vision and the encouragement of the work it took to pursue this goal. I love you both with all my heart. I am also very thankful to my friends and colleagues – Derek Haseltine, Michaela Tally, Rafael Luna, and Jelena Patrnogić – for their critical insights and feedback that made this book that much better and more accurate. The experiences and advice shared in this book would not be possible without some guidance and role models along the way. I would like to thank everyone that provided such assistance. I am blessed that the list is too long to include in full here, but I hope that you know who you are. And finally, I am appreciative of the unconditional love and support from all my parents, sisters, brothers, and extended family. Thank you, I hope I have made you all proud.

Part 1

The Beginning

Introduction

My Path – from Clueless to Clarity: Finding Strength through Struggle

I have been asked over and over why did I choose this path. I cannot say exactly when I chose it, perhaps it chose me. For as long as I can remember, it has always been something that I have done, even when I was younger, just helping others. I am not saying I was always helpful, just that I was always willing to help while also being sought out for it. It may be because I have always needed support though did not know how to consistently seek it, thus making a lot of mistakes with few initial successes. If there is a theme to my life and work, it is that there is much to learn through error and I can help make others aware to avoid the same mistakes I made. (Thereby allowing them to make new and more interesting slipups.)

I was uncertain about almost everything growing up. Going to school and playing baseball (and just about any other backyard sport) were some of the few things I could count on, mostly because I was on a team or with my friends. Regardless, this uncertainty manifested itself as mistakes large and small and none as monumental as my decision-making process to attend college. I applied very late in my senior year of high school and had already graduated by the time I was accepted to Clarion University of Pennsylvania where one of my best friends had long been accepted. Upon notification, I only had six weeks before the fall semester started and I had no plan, no classes, and worst of all, no financial aid. Ever chasing after my oldest sister, I was only the second child in my blended family to attempt college. Thankfully universities have experience with naïve families such as mine. Regardless, I essentially just showed up to campus on a Saturday, haphazardly took placement exams, blindly scheduled courses, and rashly started classes two days later.

In the fall of my second year, it dawned on me that I needed to decide on a major. Eschewing any resources or advice, I selected my major out of the university course catalogue based on three loosely related reasons: 1) I had already taken several science and math courses; 2) I had really enjoyed Basic Biology, a class designed specifically for non-science majors; and 3) "Molecular Biology/Biotechnology" sounded way cooler than regular "Biology" or the unfamiliar "Ecology." Mind you, I did not have the slightest idea what Molecular Biology was though I did know (because of the course catalogue) that it was essentially the same course of study with just a few advanced-level exceptions. So off I went to the chair of the biology department to get the paperwork signed that allowed me to officially declare my major.

Two years later after an eventually fruitful summer undergraduate research experience at Ohio University, I applied to the University of Louisville Biochemistry and Molecular Biology graduate program in the fall of 2000. (I was very busy that fall being the first in my family to graduate college while also getting married.) Attending graduate school was a passive decision as I viewed it as the next step by default. Again, not only was I a first-generation college graduate, I was the first ever in my family to consider pursuing a PhD. I had few role models and zero road maps. I did not know what else was out there for someone with a bachelor's degree in Molecular Biology/Biotechnology, especially in Western Pennsylvania. Many of my classmates were either attending medical school or pursuing doctoral degrees. One of the few active and strategic decisions I had made up to that point was choosing Louisville, where I thought I could best succeed as a graduate student. This choice was based on the fact that I had grown up in town and still had family in the area as well as the school having a small college feel despite it being the second largest university in the state.

Again, I am unable to point to a specific time that I decided to help improve the situations of my friends and colleagues. Throughout my formal academic training – as an undergraduate, as a graduate student, and as a postdoc – I was helping my peers figure out their issues, large and small. Whether it was an issue or conflict with their faculty mentor, an existential crisis, or a CV review, I was there. My friends asked me about everything. "How do I apply to a fellowship?" "What is my approach in dealing with my advisor?" "How do I reach out to a potential employer?" I do not think it was my experience or wisdom that drove them to seek out my counsel back then; maybe it was my ability to appear calm with a down-to-earth demeanor. However, beneath the surface, I was really in no better shape than they were as I very much needed the same questions answered for myself. I have always been a people person, and it was through helping others that I found some solutions for my own problems. Like I said, I have made so many mistakes and continue to do so. Hopefully though, I have learned from some of them. Compared to my classes and research (the things I was supposed to be doing), it was very fulfilling knowing at the end of the day I helped a colleague or a group of peers.

In the spring semester of my second year of graduate school, I was struggling mightily as I failed my qualifying exam. My proposal was so poorly composed I did not even get the opportunity to defend it. I was placed on a remedial track where I had to take an additional course in the fall (Molecular Biology), rewrite my proposal, and have regular check-ins with the graduate program director. Having been humiliated and delaying my progress by another semester, I finally passed the qualifier and reached PhD candidacy.

I chose my PhD lab because I wanted to do proteomics research without really knowing what that meant. However, my dissertation project did not involve proteomics as I had come to understand the field. My project entailed dosing kidney cells with high glucose and characterizing the subsequent kinase and metabolic signaling cascade that was activated. However,

Introduction 5

I continued to struggle as I dove deeper into my research project. I would maybe do one successful experiment a week, maybe one a month, that produced any data worth sharing. It was not enough to keep me going, to keep me motivated, and I knew I had to finish my research somehow. I was waiting for help instead of asking for it. I expected my advisor to rescue me though I never solicited his advice. It got so bad that we had a serious conversation about whether or not a master's degree was even still possible. This is the part of obtaining a PhD they do not tell you about. You are creating new knowledge and no one, not even your advisor, knows for sure if your research project will work. I only learned this cruel lesson after the fact. I was honestly working hard, but my experiments were hardly working. To make matters worse, I was comparing my progress to another grad student who seemed light years ahead of me. I was lost in the wilderness and had no clue. Then after struggling some more, wandering in the dark, and doubting myself, I made a huge breakthrough in my project. Basically, the alternative signaling pathway that I had been pursuing instead of listening to my advisor was essentially proven. Everything and nothing changed after that.

In addition to figuring out my research, I paid more and more attention to my desire and interest to help develop careers. Honestly though, it was mostly in my own self-interest since I also needed to develop my own career. I realized that if I had an issue, then my classmates probably had that same or similar issue. I began exploring more how to do that, talking with faculty and anyone else that would listen. Thankfully, there were a few very sympathetic faculty that told me there was more out there than just being faculty and doing academic research. I became involved in organizing a student-led group for the biochemistry department and was elected as its chair within a year. I was included in recruiting incoming students and was appointed to the Graduate Executive Committee. I was just trying to develop myself and in doing so, began to help develop others. I organically formed a team around myself where I was interacting with very smart people who were helping me figure out my research issues and I was, in turn, learning how to help them with their career or professional development. This was a symbiotic reciprocal arrangement that taught me how to delegate responsibilities and manage my peers. It was really fulfilling to do. I found that there was a tendency that my assistance was helpful rather than harmful. If I ever did give bad advice, and I am sure I did, they were either courteous enough to not let me know or smart enough to ask someone else the next time.

After making steady headway in my project and without any input, I made the decision that I was ready to write up and defend my dissertation. To the point that I even launched my postdoc search after presenting a poster at a national conference in the spring of 2006. I did not ask permission, nor did I seek advice. I applied widely and since I had in-laws that lived in Frederick, MD, I focused on NIH labs, especially those on the NCI-Frederick/Ft. Detrick campus. Using the NCI central postdoc portal, I had a few PIs interested in me. Surprisingly, I landed an interview with a genomics lab, a completely new area of research for me and got an offer to be Cancer Research Training Fellow. I was not alone on the job search, my wife, fully expecting me to have finished my PhD, also landed a job in Frederick as a teacher. However, I did all of this before ever giving my dissertation committee an update on my research progress. Much to my dismay and in hindsight not a surprise, my dissertation committee, in full agreement with my PhD advisor, determined that I was not even close to being ready. Having made a huge miscalculation, I, embarrassed and very much humbled (again), had to ask for a deferment of the postdoc job offer for at least a year. My wife not only had to do the same but also had to ask for her old job back.

With renewed commitment and a focus that I had not thought possible, I generated more data in the subsequent nine months than I had in the previous five years combined. In the end, less than a year's work constituted the bulk of my dissertation. By March 2007, I had permission to write up my dissertation and make my PhD defense. Three months later I successfully defended. Because of the previous year's false start, I made another series of huge mistakes. I moved my family to Frederick and began my new postdoc fellowship (in a new research area, remember) only a week after my dissertation defense. I started my new position before my degree was conferred; thus, I was not yet eligible to receive the full postdoc stipend. Worse yet, I left my PhD lab before I had converted my dissertation into a manuscript for submission. It took me another 12 months of back and forth with my graduate advisor before I realized that I would never submit, much less publish, a first-author paper based on my PhD work. The loss of this opportunity furthered my desperation to find a career away from the bench. However, I was surprised that others in my former lab had continued to expand the central observation of my research in new ways, ironically by doing a proteomic analysis of high glucose–induced signaling in kidney cells. Eight years after I graduated, I was included as a second author of a paper based on my original PhD work.

When reflecting on my postdoc experience, among the many challenges, I remember a few moments of success and satisfaction. The most impactful is the eventual realization of my career trajectory. The most lasting is the circle of friends and colleagues with which I surrounded myself and am still in contact. And the most satisfying, perhaps, was winning the Ft. Detrick Fun League Softball Championship. Let me explain. Since I joined my postdoc lab in the summer, it was natural that the lab PI shared that he was on a softball team that played on base. I expressed my interest in joining as I had been playing sports my entire life and wanted a competitive outlet as well as an excuse to network with and impress my boss outside the lab. It did not go how I had hoped. He left the team in 2008, after only one season playing with him. Mind you, he had probably been on that team for a decade. I doubt I was the reason he no longer played, but we never talked about softball after that. Even though it felt like a missed opportunity, it was also still a chance to expand my network beyond the lab. In my third and final season, my team tore through the league with only a single loss and in so doing, we won the regular season and tournament championships. This was the pinnacle of my postdoc tenure: a Fun League softball team triumph that had nothing to do with research, publishing, or finding a job. While I was playing softball, there was also a lot happening in my postdoc training.

As I have stated, helping others, outreach, and service were very satisfying. I moved from a graduate student helping graduate students to a postdoc helping fellow postdocs. It was infinitely more rewarding than doing research. I began to strategically seek out more opportunities that followed my interests, sometimes at the cost of my research projects. As a postdoc, I was drawn to and eventually played a major role in the Fellows and Young Investigators (FYI) Steering Committee, the equivalent of a postdoc association at the NCI. This was a seminal and formative experience for me. During my tenure, I was elected as chair and also led the scientific sub-committee. Additionally, I ran the campus-wide postdoc seminar series and new postdoc orientation. I worked on the postdoc training satisfaction survey and, because of my growing reputation, was invited to represent trainees on several committees across Ft. Detrick and the NCI. I sought out teaching and outreach opportunities that led me to lecturing high school and college students on poster presentations and cancer metabolism. Even when I realized that what I was doing informally and for fun might be a viable career path, I still did not fully understand the idea of managing or directing graduate or postdoc affairs. From this point, I became very focused and strategic in what I chose to pursue. I began to elevate how I communicated and framed my experience to respected leaders in the field. It was an important shift in thinking for me because now I had a career target to aim my efforts toward. Because of this new clarity, I was able to transform my experiences into an internship with the NCI Office of Training and Education. Regrettably, many of us do not have this frameshift about our next job and how to get there until it is almost too late. It was around this time that I began to see the fruits of my many non-research labors. I was blown away to find that my fellow postdocs were now actively seeking me out to expand their own professional networks.

Unfortunately, the lack of progress in my research projects caught up to me. At the end of my second year of postdoc, I found out from a department administrator, not my PI, that my contract was not going to be renewed. As a result of this miscommunication (with the administrator, not the termination), I was finally informed as to who my primary mentor actually was. It was not in fact the guy I tried playing softball with, it was his very experienced staff scientist. It was not like I did not know the staff scientist, we worked closely together, but I always wondered why he was telling me what to do in my projects and inviting me to attend meetings with him. Not until two years into my postdoc did I realize this person was my assigned mentor. Regardless, thanks to the generous termination policy at the NCI/NIH, I had 12 months to find a new position. I wasted no time and was still quite uncertain about my next steps. So, with the help of a lab mate's introduction, I made a solid connection that led to an interview for a second postdoc. At the interview, I used slides from my PhD defense seminar, not my postdoctoral work. There were two really good reasons for this. First, the lab I was interviewing with was hardcore cancer metabolism, and my dissertation could be re-framed as being metabolism related. And second, I had not generated a shred of usable data from my current postdoc. Even though it was obvious to both of us that we were an excellent fit for each other, I was initially turned down. As you can imagine, the combination of being fired, turned down for another job, and not knowing if I could pay my bills led to extreme anxiety and panic. Humbled (yet again) and very desperate, I reached out and asked to be reconsidered by the metabolism group. I was eventually hired into that lab and found out later that two of my future lab mates had vouched for me. Interestingly, one of them was the third baseman on my softball team. I was able to return the favor to the second as I helped them to transition to job in industry 10 years later. As stated, this postdoc was a much better fit for me both scientifically and personally. I was fully committed to the science and also negotiated that I be allowed to continue my internship and related training affairs work outside the lab. For the next 18 months, I flourished. I felt I had positively impacted the lab and also had the honor of being specifically named in my research division's successful program review for my contributions to the greater NCI-Frederick community.

Despite my efforts, to this point, I had zero research papers to my name. I had already come to terms with the reality that a future doing research was not possible. The next step in my journey to creating a new career trajectory was actually convincing myself (*and then my wife*) that it was worth all the hardship and anguish of getting my PhD, doing multiple postdocs, and moving my family across the country twice. To help make this transition, I took advantage of the resources around me at the NCI as well as the NIH. I sought out role models and training workshops. I even made friends with the director and staff of the NIH Office of Intramural Training and Education. Perhaps most helpful was meeting with a career coach and taking the MBTI and DiSC self-assessments. This effort along with some much-needed self-reflection provided the necessary tools to recalibrate what it meant to be a PhD and a postdoc. I had to reprioritize the skills I possessed and the skills I needed. I would be leveraging all of my experiences as a scientist and applying them to a completely different set of problems. I would no longer be doing research but transitioning into a research-adjacent position where I would be helping people deal with issues that hindered their research and their professional progress. I realized that it will never be me saying, "I am going to cure cancer," though maybe I can help someone else realize their full potential, by raising their awareness of or even removing some barriers that might be harmful to their research. I would now be saying "I coached that person and they made their situation better allowing them to take major strides toward a cancer cure." This was a significant and necessary change in perspective that allowed me to believe this was the right path. I understood that this could be enjoyable while also being important. Now all that I had to do was convince someone to pay me to do this.

However, it was difficult finding a job, especially one I was excited about. Though I had several interviews for similar positions, it took many more months of gaining additional experience and training outside of my lab responsibilities to qualify for my current position. I even went so far as to pitch the graduate dean at Louisville on the value of having a postdoc office and a professional development training structure in place in the hopes of them hiring me. It almost worked. I would not have my current position if I did not participate in the FYI Steering Committee. Not only did it give me a glimpse into an administrative career, it helped me tap into a vast network of people in the field. I was able to hone my leadership skills – learning how to manage meetings and delegate assignments across a large group of volunteers. In particular, it was my internship in the training office and the subsequent administrative understanding that made me a top choice for the position. I would not have even known about the position if it were not for a few people who knew of my career interests sending me the job ad.

After four years of postdoc that included two years of job searching, I started my position as the director of the office for postdoctoral fellows at Harvard Medical School in June 2011. They were a little hesitant about hiring me straight out of my postdoc though I was also able to connect with and support the postdocs at HMS on a deeper level than they had previously experienced. I did miss the scientific chatter of the lab as well as my lab mates; however, I never missed the day-to-day frustration of failed experiments. I definitely felt that my "Eureka!" moments happened a lot more often after I transitioned away from the bench.

As Director of the HMS/HSDM Office for Postdoctoral Fellows, it still took several more years of professional development to gain trust, build relationships, and find my footing. I now manage a vibrant office of two that serves thousands of postdocs. I am responsible for developing and executing programming that adds value to the postdoc training experience. There is no typical workday as week-to-week I am alternately presenting or hosting programs for postdocs while juggling the administration of the office. One of the first tasks I had was enhancing the office's visibility, use, and reputation not only within HMS and Harvard community but also locally and then nationally. My office currently produces over one hundred events, seminars, and workshops per year that cover postdoc and research skills, as well as career, professional, and individual development. Since my arrival at HMS, I developed and instituted a flexible curriculum predicated on creating awareness, building skills, and establishing experience to help address major postdoc pain points of research expertise, publication record, funding stability, and career readiness.

Having been a postdoc and now running a postdoc office, I have witnessed trainees struggle with identifying a career path, building an identity and vision, and receiving satisfactory mentorship. Relatedly, some of the biggest challenges I face in my role are finding interesting ways to engage postdocs on these topics while encouraging them to develop themselves and get into the right frame of mind. I advise and coach trainees on personal and professional matters including career exploration, networking, job searching, application strategies, and interviewing. I counsel postdocs that their research skills are highly valuable and should be used in every endeavor of their scientific and career progression regardless of its relation to research. Much to my satisfaction, I have been invited to speak and write about these topics across the country and the media. The most demanding aspect of my current role is advocating for and interpreting reasonable and fair postdoc policies and guidelines at HMS. I am actively involved in developing and implementing procedures for orientation, onboarding and exit, benefits, compensation, term limits, and individual development plans (IDPs) for my fellows.

While I have applied and interviewed for several positions since, being hired at HMS has been my only successful professional job search. This is not the end of my career story, and I do not know what is next on my horizon. Although there is uncertainty, it is no longer joined by the terror that used to be its partner. Upon reflection, when I faced uncertainty in the past, I built teams around myself while learning to overcome the fear of asking for help. I then began to frame my mistakes as growth opportunities instead of failures. Ultimately, through hard-won experience and often-disregarded advice, I was able to revise my approach to career and professional development to leverage the skills I earned as a scientist. I am no longer driven by panic or anxiety. I am in control of my trajectory and will now move forward by following my curiosity and joy.

1

How to Use This book ...

Are You Puzzled?

You probably have at least one issue, if not many, that you are puzzled about. That is why you are reading this book. I can anticipate most of your struggles as they run the full gamut of making the most of your postdoc. I am certain you are asking: Is there a postdoc protocol I can follow in framing my non-linear path? How do I gain and practice situational awareness? How do I navigate my way through a postdoc? What is my path to independence? What is career transition readiness? Can I get some advice on my application materials? How do I prepare for interviews and job negotiations? How do I develop an exit strategy?

I may not be able to answer every question you bring with you as you read, but what we are covering in this book cuts to the heart of these major questions. When leveraging your PhD or research training experience, you need to be very mindful and active. Research progress, career advancement, and professional development are not passive processes. It involves some self-assessment, convincing yourself of your value, and persuading others that you are worth investing in and eventually hiring. This book lays out strategies to address the struggles listed above, but I warn you that it is not comprehensive. Moving from beginning to end, you will learn to start with the finish point in mind. You will become aware of pivot points, resilience, and expectations. You will understand that plans lead to priorities and those lead to the next steps of launching your career. Finally, you will learn how to create the application materials and tell the stories that will lead you to your postdoc exit point.

While most of the experiences and examples that are shared come from a primarily US, academic, and biomedical life sciences perspective, I believe this book can be applicable to all science-related trainees including international and industry postdocs. There is a certain universality in postdoctoral training that crosses all demographics and backgrounds, and I hope that this book rings true to your own experience.

DOI: 10.1201/9781003285458-3

How to Use This Book If You Are a Postdoc

Dear Current or Future Postdoc Fellow,

You are faced with many obstacles in your path to independence, none more frustrating than getting the most out of your training. I wrote this book to provide a protocol of sorts for implementing strategies in choosing the right research environment to thrive as a postdoc as well as planning, and executing, a successful postdoc tenure. Written for current or future postdocs, the content will cover what I believe you need to know, and do, to efficiently advance in your early research career.

However, very little goes according to plan. Regardless, you can use this book as a guide that I never had. In my transition to and eventual training as a postdoc, there were so many unknowns that almost nothing went according to plan for me. To be honest, while I may have enjoyed the journey, I did very little to plan my trip. Everything I did was new to me. I have claimed to have made every mistake there is to make in my academic training.

My advice is to read this book in any order or sequence you choose. I have separated the sections into Beginning, Middle, and End. Not so much as to ensure you digest it chronologically but to break down the sometimes indistinct phases of postdoc training. I have no idea where (or when) you are in your timeline, but I do know there are certain things you need to know at each step. Use this book as a measuring stick to help you establish benchmarks and milestones along your journey.

I am so thankful that not only are you taking the time to read this book but, more importantly, you are seeking advice and resources to help you in your career and professional development. You now have the difficult task of taking in information, parsing its relevance, and either keeping or discarding what you have learned. This is a fundamental part of the scientific method, and the more data you collect, the better your "experimental design" will be. Of course, the hypothesis that this grand experiment will test is whether or not you and your career are properly prepared. The next step is to take action and begin using and implementing the new knowledge to develop yourself.

My hope is that in doing so, you set yourself on a path toward independence and eventually to your authentic self. This book is also meant to dispel myths, shine a light in the darkness, and act as a reality check for you. If you are holding this book in your hands right now, I imagine you have a million questions and even more doubts. While I do my best to address them, you can apply the concepts you learn to build a framework that starts and sustains important conversations. As you continue reading, you will gain the necessary confidence to meet your career and professional development challenges head-on.

How to Use This Book If You Are a Faculty Mentor

Dear Faculty Mentor,

For you, the appointment of a postdoctoral fellow initiates a temporary and defined period of advanced mentored training. Your postdoc fellow is expected to engage in independent research while also developing skills and experiences that support their specific career goals and professional development. Your role as their faculty mentor is critically important for your postdoc's training experience. Your mentoring relationship is an alliance built on bi-directional respect, open-mindedness, frequent communication, and a willingness to adapt as needed. To that end, this book outlines guiding principles for your early and ongoing training discussions. You and your postdoc are encouraged to return to these pages as progress is made. This can be especially helpful when preparing for larger planning discussions or important training transition points.

While primarily written for postdocs from the perspective of a former fellow and current postdoc affairs professional, there is much you can gather from and add to this book. It is very important to remember your postdoc fellow chose to learn from you and your research program. You should not overlook or take for granted the trust they are placing in you. Also, do not discount the value of your own academic experience (from grad school to postdoc to faculty) when coaching and counseling your trainees. Though the times and how we view research training have changed since you were in their shoes, your knowledge, advice, insight, and guidance are as relevant as ever.

If I may be so bold, I would like to offer you some advice and share with you the wisdom I have gleaned from postdocs. You need to offer them a path to independence with a thoughtful plan and explicit expectations. You need to give them room to grow across many domains including permission to make mistakes, to pursue their interests, to develop themselves, to be human, to be vulnerable, to be brave, and to take risks. You must be proactive in your mentoring while providing resources and multiple perspectives inside and outside of your lab. Please have open, transparent, and ongoing conversations that cover more than just the next set of experiments. Show them you care about them beyond what they can contribute scientifically. You can read this book and, agree or not, you can share your own wisdom from the lessons you have learned throughout your career.

How to Use This Book If You Are a Postdoc Affairs or Career Office Leader

Dear Training and Career Offices,

    A trainee's drive to successfully develop themselves should be guiding your programming to be flexible and adaptable across the depth and width of your curriculum. Raising awareness in career-related topics is a universal and inclusive approach for all trainees, which affords the opportunity to openly explore diverse career options. Training that builds specific professional skills provides trainees with low-risk, high-reward opportunities. Although experiential programming is the height of strategic career and professional development, it is also the most time-intensive and selective. Regardless, your implementation strategy must emphasize discernment by aligning coaching and programming needs with postdoc concerns and practical application of their scientific training. When postdocs and PhDs pursue career and professional development, they are usually seeking meaning in their training by enriching their learning and service experience, improving their relationships, and enhancing their job preparedness. With increased awareness, skills, and experience, trainees are able to go out into the professional world fully prepared to make evidence-based decisions about their desired career trajectory.

    Your offices should be providing resources for transition and decision point support as these are the most anxiety-inducing moments of their training outside of research progress and publications. They are also the topics to which they pay the least attention. You should then create the bulk of your discussions or workshops around these topics. You can help them to acknowledge that life and lab are messy. As soon as they accept and embrace that fact, they can move on to the important work of self-development. You also need to help them see that change is inevitable and begin to view it as an opportunity for growth and not something to be feared.

    I recommend approaching this in a stepwise manner to incorporate its principles into your training curriculum. The contents of this book can bolster your office's strengths while also identifying areas of innovation. You can then deliver the appropriate training opportunities when needed. It is important that in your role you develop and reinforce both a clear training path to independence and attractive educational exit points. In addition to this book, there should be infrastructure in place to help postdocs accomplish this, which can be enhanced by establishing standard mentoring practices and expanded use of institutional and external resources.

How to Use This Book to Improve the Research Enterprise

The success of a postdoc fellowship, and the scientific enterprise as a whole, depends on active trainee, faculty, and institutional investment in the development of each fellow. The ideal outcome for this is a better research product, retention of committed trainees, decreased time to completion, and improved workforce readiness. Along with an increased visibility of the research produced by your trainees and alumni, the ultimate payoff is growth and improved quality of your recruitment pool. Taken together, this means that your trainees will eventually be more competitive at the institutional, local, regional, and national levels for funding and job opportunities. I wish that one day we can overcome the "do more with less" mentality that pervades graduate and postdoctoral training and actually "do more with more." To that end, I hope that this book will be used to continue the evolution of postdoc training toward its next phase.

2

The Most Important Professional Decision You Will Make … So Far

In early 2008, I was asked by some former graduate school colleagues to share my experiences leading up to landing my first postdoc position. You would not be surprised to know that I readily and happily gave my opinion on the process. What is surprising, however, is how well it has held up. In the years since, I only had to update a few things that I learned along the way. In fact, due to the glacial pace of training evolution, I find that I am still giving this same guidance even today in my current role.

In July of 2006, I accepted a postdoc position at the National Cancer Institute in Frederick, MD. A year later, I was finally able to begin my postdoc after those grueling months of running just "one more" experiment for my dissertation. I quickly found that my responsibilities as a new postdoc were very similar to graduate school: read, do not break anything, and ask "someone else" if you cannot find what you need. Yet, I also realized that I had greater ownership of my projects and was encouraged and expected to explore my own hypotheses. I am still amazed at how long it took to adapt to my new lab environment. The elation of writing and defending my dissertation had definitely worn off.

To help those of you looking for your own postdoc, I will cover many of the things that I did, or should have done, in landing a promising postdoc after grad school as well as sharing my experience and reflections from advising postdocs for over a decade. I will add two more pearls of wisdom and these may be more important than anything else I might say – enjoy yourself and be visible.

To Postdoc or Not to Postdoc

Postdoc Position

Many PhDs (and MDs) choose to extend their academic research training while gaining new technical and professional skills by pursuing a postdoctoral research experience. Furthermore, you may feel compelled to select this path because pursuing a postdoc has become the *de facto* requisite next step to most careers in research and biomedical science. Regardless of your motive(s), there are many practical advantages to the postdoc experience: a path to independence, a chance to publish, a reputation to build, and a career to cultivate. Admittedly, there is risk and reward in every career decision, but there are particular costs to doing a postdoc that could have real and potentially lasting personal, financial, and professional ramifications. Long gone are the days when you could graduate with a PhD and be hired directly into a faculty position. There are rare but increasing opportunities outside of the faculty realm like industry, the private sector, and higher ed, to make the leap from graduate school into the workforce.

Postdoc Skills Outlook

In recent years, several trends have emerged that could significantly impact the future of postdoc training such as trainee unionization, living wage considerations for salary, expansion of professional staff scientists in academia, and specialized career-focused postdoc fellowships. Regardless, postdoc training remains a virtually required step for many PhD positions in the sciences as part of the current research workforce model. With this in mind, you should be considering the skills necessary for successful career placement and longevity from the very start. For example, asking yourself in-depth and searching questions before embarking on a postdoc search (or job search in general) allows you to strategically pursue a more focused opportunity. You need to understand which skills and attributes will set you apart on your desired career track. You should consider the benefit of cross-training in certain skills that will give you the "most bang for your buck" in a variety of career paths. After reflecting on these, you and your faculty mentor should figure out how to strategically build those vital skill sets within and outside the lab.

Ideally, this skills identification process should have begun in graduate school. However, I can attest from my experience working with postdocs that this is not often the case. Individuals with lengthy postdoctoral experience repeatedly ask me these very questions. Thankfully, the National Postdoctoral Association has outlined the six core competencies of postdoc training. These recommendations provide clarity as to what constitutes a successfully trained fellow. These competencies can be categorized into *Research Skills* and *Transferable Skills*. Through combined and directed efforts between you and your faculty mentor, training in research skills covering in-depth knowledge in a discipline, principles of laboratory experimentation, and responsible conduct of research is by and far successful. Transferable skills, on the other hand, are usually poorly addressed in the lab and tend to be left up to you to learn and manage. Everything being equal, effective communication, professionalism, leadership, and management are the types of transferable skills that will set you apart from your peers and, let us say, your competitors. Early assessment and implementation of these core competencies give you an advantage not only in navigating your training but also in choosing the right environment in which to pursue your research.

Postdoc Job Search

The Postdoc Search

With few exceptions, the entirety of the postdoc hiring process is the sole responsibility of the hiring faculty and their staff, including but not limited to advertising, recruitment, applicant tracking, interviewing, and onboarding. As you can imagine, this usually leads to a passive, email-based process that relies on labs with open postdoc positions depending on qualified candidates reaching out on their own. On top of this porous practice, many labs do not even advertise. To complicate matters more, the postdoc job search is unique in its cycle and unlike graduate school recruitment or academic faculty hiring. While PhDs have distinct graduation and degree conferment dates (typically May, August, and December) you can be hired on as a postdoc anytime throughout the academic year.

It may be necessary to stay on for a few months after your dissertation defense to wrap up your project and hopefully publish your work. However, it is not ideal that you do your postdoc in the same lab, or even institution, that you received your PhD. There is a sense across many disciplines that if you do not move on to a new training environment, then you may end up lacking a path to true scientific independence. In addition, many institutions have a culture of not hiring their own trainees while creating an "up and out" training pipeline. If you are an international trainee, you may face pressures, expectations, and/or lack of opportunities in your home country that virtually require you to pursue postdoc training abroad (in this case, the US). The same may be also true for those that received a PhD in the US and do not have postdoc opportunities at home. Because of these practices, it is never too early to explore positions that interest you. Do not be afraid to contact anyone and do not be offended if they never respond. While you should definitely engage your PhD advisor in your postdoc search, do not depend on them to find you a job.

During this process, you should be establishing a timeline based on when you can realistically be finished with your training. Ideally, this would be at least 6–18 months. You can then work backward from there to build out a plan wherein you, among other things, keep up with current literature both within and outside your research area, create automated job searches, and attend meetings and conferences. In addition, you can find potential postdoc opportunities by perusing NIH RePORTER to find currently or newly funded labs in your field. When faculty do advertise postdoc openings, they tend to post them on their own lab website, Twitter, or LinkedIn, or sometimes on postdoc-focused job sites. Recently, there has been a growing use of specific postdoc recruiting events where faculty work with their institutional postdoc offices to capitalize on centralizing the process. When possible, you should submit an application for these larger open recruitment events. Chapter 9 provides extensive instructions on crafting your application materials.

The Postdoc Mentor

Since you are reading this book, you have likely already chosen the postdoc path. Having done so, you are now responsible for *THE most important decision* you will make as an early career scientist: with whom you will train. You must recognize that the postdoc position, more than any other position before or after, will be solely dependent on your PI, or as I prefer to call them, your faculty mentor, with little or no departmental or institutional oversight. The relationship with your faculty mentor is one of the most important factors that will make the difference not only in the rate of your training progress but also in the viability of your eventual job search. You should also consider whether you will best fit or succeed in a larger or smaller lab as well as the relative career stage of the PI. These could play a significant factor in how much access and interaction you may have with them along with where and how often you publish.

Your postdoc faculty mentor, as well as your graduate advisor, exerts a vast amount of influence on your career. Whether you are in graduate school or pursuing a postdoc, you need to be cultivating a well-rounded relationship with your faculty mentor. For one, they are going to be a necessary and expected reference for you for several years and positions to come. The success of your postdoc training, and your career beyond it, is as much about vetting a faculty mentor as it is about aligning research interests. International scholars must pay special attention to this point. Your relationship with your faculty mentor as well as your research progress are closely tied to your Visa status and any conflict or tension can be exacerbated by this.

The Postdoc Interview

If you are armed with any semblance of a career trajectory along with knowledge of its requisite skills, not only would you be ahead of the competition, but you could also be very strategic in probing how well you may fit with the faculty mentor and their lab. Chapter 9 offers a full overview of interview preparation from initial contact to offer. When contemplating a postdoc position, you should consider compatibility in mentoring styles, training expectations, and research interests. You should then inquire about the security of the funded position, access to facilities, and the lab's recent track record of publishing. Do not miss an opportunity to meet with current or former lab personnel to further assess the lab culture as well as the career outcomes of lab alumni.

Upon securing a postdoc interview, you need to begin doing your homework on your potential landing spot. Of course, not only should you be reading their recent publications, but you need to also know their major contributions to science. It also helps for you to have an understanding of the history of the field. When preparing for your interview presentation, you will need to craft a compelling story that covers a cohesive thread of discovery. Make sure you practice your presentation in front of several different audiences. Incorporate their feedback then practice your talk again (and again).

As you are having your interview conversation, be honest in your responses by taking credit where it is due, share what skills you bring to the lab, and admit what you still need to learn. Remember, however, you are also interviewing them, therefore ask a lot of questions. Scientists love to speak about their own work so let them talk. When interviewing, you must also be aware of potential "red flags" that may warn against joining as well as "green flags" that may affirm toward a good fit. For instance, a red flag may be that you do not meet with everyone in a lab or you talk to others only in the presence of the PI. Conversely, a green flag may be that you have an off-campus lunch with all the members of the lab and the PI stays behind to give the group a safe space to share.

Some questions to ask that will give you a better sense of mentoring style and lab culture:

- How long is this project funded? Will I need to generate additional funding?
- What are your expectations for me in this postdoc?
- What have your recent or successful trainees gone on to do?
- What is your mentoring or management style? How will I be evaluated?
- How are projects allocated?
- How are manuscripts written? Where are they usually submitted?
- How will you foster my independence?
- What piece or parts of my project could I take with me?

For internationals, your immigration status affects almost every aspect of your existence in the States and the first question above is important to ask since you are likely not eligible for many independent federal postdoc fellowships or grants. For all potential postdocs, the last two questions hint toward your postdoc endpoint. This is where the real talk of your training starts. You will be able to gauge whether your potential faculty mentor actually has a plan for their trainees and you in particular. This is especially important when choosing whether or not to pursue your own faculty search. I am not saying, however, that taking a less planned out "wait-and-see" approach is bad, far from it. I just caution you that this approach leaves a lot of decisions open ended and out of your control.

The PhD-Postdoc Transition

Once you accept an offer for your postdoc fellowship training, you still need to deal with and determine many more things. First of all, congrats, you now have two bosses! You will need to negotiate a start date (and maybe salary) with your new postdoc mentor while negotiating your exit with your current PhD advisor. Among other things, you will need to discuss how best to wrap up and hand off your graduate project(s) as well as attending to the details of your graduation requirements. Perhaps most importantly, this PhD exit conversation should include the next steps for ensuring that any outstanding manuscripts are finalized and submitted with agreed-upon authorship. It is vital that you make every effort to at least submit your PhD papers, possibly even seeing to their acceptance, before leaving your graduate lab. Many trainees choose to stay on for a few months after their defense before starting their postdoc fellowship for this very reason.

The Postdoc Location

There are two methods to targeting your postdoc search that I recommend. You can focus on the who (i.e., faculty mentor) or you can concentrate on the where (i.e., institution and/or location). Not far behind the decision of who you will work with is the important question of where that will take place. It is not just the actual institution that will support your research but also the physical environment in which you will live that is an important consideration. Just as in real estate, aspects of your postdoc success and wellbeing can be charted to your location, location, location. As you are interviewing, you should be scouting out reliable sources of information (i.e., guidebooks, websites, postdoc offices and associations, and future peers and colleagues) about the surrounding area. It is necessary to know where you may be moving and what it realistically costs to live there. You should include cost of living in your job offer calculations.

> During my career, I moved 3 times in 11 years for work and found that the cost of living increased for each new city. I went from living with my parents for free outside Pittsburgh, PA to paying $450 per month for a 2-bed, 1.5-bath apartment in Louisville, KY to paying a $1500 mortgage plus escrow for a 3-bed, 2.5-bath townhouse in Frederick, MD to now paying $2000 for a 2-bed, 1-bath apartment outside Boston, MA. While my salary also increased at each step, the relative cost of living in the new city virtually nullified those gains, at least initially.

In light of the cost of living, some of you might have to decide between a renowned lab in a high-cost-of-living area where you are scraping by and a less well-known lab in a low-cost-of-living area where you live in relative abundance. Furthermore, since it can be quite expensive to move, be sure to ask about relocation reimbursements or allowances (or even a lease guarantee), as it is not always a standard practice for the labs to offer. For instance, if you accepted a postdoc (or any job) in the Boston metropolitan area and leased a place to live, you could expect to shell out nearly $10,000 to make that move. In addition to the mind-numbing logistics of moving, some of the other things you need to be thinking about are the locations of good schools and the most efficient commuting routes. Ideally, you can find a place to live as close to the lab as possible because late nights and weekend projects are easier to handle. However, this might not be feasible due to a number of reasons, some of which are outlined above.

This is not to mention the very real culture shock that many international (and some domestic) postdocs feel as you move to or across the US. As you may well find out, the US is huge, and it varies culturally across its many regions, states, and cities. Since you are pursuing a postdoc in academia, you will likely be in a relatively welcoming research institution that contains and has experience with other internationals such as yourself. But these academic enclaves of international awareness can be surrounded by local residents unfamiliar with the global reach of the research enterprise. Nevertheless, you must also navigate the new logistics of living in the US. You will need to find housing (as outlined above), sometimes from abroad, deal with the exchange rate, and possibly learn how to pay your taxes. You will likewise be introduced to a new and potentially very confusing system of healthcare management and insurance. And you will deal with all of these issues as non-native speakers (and writers) of English. There are many resources available within your new institution that can help you manage your onboarding including your department administrators, postdoc office or association, human resources, and the international office to name a few.

Creating a Productive Environment

When it comes to postdoc training, no one plans to have conflicts or differences in opinions with their mentors or lab mates. However, it is crucial to your success (and theirs) that you intentionally create a productive postdoc training environment predicated on mutual respect and professionalism. It is reasonable to expect that your faculty mentor will always treat you professionally and respect your individuality, which includes recognizing that your goals and professional values may differ from their own. Faculty mentors should value and consider your opinions while delivering feedback in a respectful and constructive manner. At the same time, you are also expected to be open to feedback and respectful of your mentor's time and opinions while making sincere efforts to learn and move research projects forward. Furthermore, both you and your faculty mentor are responsible for the continuation of your training in the rigorous ethical conduct of science.

Once the faculty mentor and training environment choice is made, you must discuss and choose a promising research direction. This project will set the tone for the next several years while providing myriad development opportunities. In considering a timely and feasible primary project, you need to steadily move the project forward while building productive collaborations. Also, breaking down your research into smaller, and hopefully publishable, pieces alongside a solid secondary or tertiary project will allow you to regularly demonstrate productivity. Additionally, having working secondary or tertiary projects gives you a chance to explore more science, a place to fall back on when your primary project is completed or stalls, and a natural independent research direction to take with you upon your departure.

It is easy for you to fall into the trap of thinking that publications are the only way to show and validate your research output. I would argue that communicating productivity takes many forms such as lab meetings, mentoring, grant writing, talks, posters, journal clubs, preprints, patents, networking, and much more. Since each of the above is an opportunity to share and demonstrate your postdoc progress, you should place emphasis on making sure there is a support infrastructure surrounding you that offers this type of training, such as a postdoc office or an association. The major priorities of your postdoc training should be to gain independence, build a professional identity, and create a vision for the future. Whereas the ultimate goal is to secure a job that you are happy with. These are all key indicators of your success and though you are responsible for the bulk of it, you are not alone. You have many investors and stakeholders in your success including your faculty mentor, your institution, any agency that provided funding, your future employer(s), and of course, you.

3

The Postdoc Protocol

Postdoc Process

I remember my teachers and parents asking me as a kid, "What do you want to do when you grow up?" I sometimes still ponder that question today. While it is an important one to consider and, as PhDs and postdocs, you need to better understand where you are headed. Having been a higher education professional, two-time postdoc, and a grad student, I know first-hand that the process of scholarship, research, career advancement, and professional development are neither quick nor passive. Instead, you should actively approach your career development using a methodical process of self-knowledge and reflection by envisioning an endpoint, assessing your situation, developing plans, accomplishing goals, telling stories, maintaining progress, and creating your exit.

In developing your own postdoc protocol, you start your training with the end in mind. Thus, you are much better equipped to set expectations and strategically position yourselves to take advantage of growth opportunities. You can then identify pivot points while avoiding pitfalls through heightened self- and situational awareness. This process also encourages you to plan for research and training challenges with the guidance of your mentor. When you are encouraged to develop and share your career story, it compels others to be your advocate. Ultimately, you can efficiently proceed toward your desired endpoint by combining research, professional development, and career advancement efforts.

Envisioning the Endpoint

Defining Success

Before the postdoc process begins, you must define what makes a successful postdoc. When asked to explain this, my go-to response is a successful trainee is one who has obtained a job. That may be overly simplistic, but the definition of success depends upon whom you ask. For instance, I pose this very question to postdocs at every individual development plan workshop I run. They respond with a lengthy and wide-ranging list that goes from networking to publications but interestingly, "gaining independence" is deemed the most important success factor. When asking a group of faculty mentors, however, they all agreed without hesitation that a successful postdoc depended almost entirely on publications. While I am not surprised by either group's opinions, I remain astonished at the huge disconnect in outcomes.

Perhaps the most straightforward description of a successful postdoc can be found by merging their respective definitions: *Successful* means accomplishing an aim or purpose while *Postdoctoral Fellow*, according to the National Institutes of Health and the National Science Foundation (NIH and NSF, respectively), is described as an individual who has received a doctoral degree and is engaged in a temporary and defined period of mentored advanced training to enhance the professional skills and research independence needed to pursue his or her chosen career path. Consequently, a *Successful Postdoc* can be defined as an independent research professional who has pursued his or her chosen career path. With few caveats, most researchers would agree that this definition is a great starting point.

Create a Postdoc Trajectory

While it is vital to define the parameters of success, it is much more difficult to achieve it. The sooner you take the time to explore your career trajectory the better informed you will be during your training. There is a feeling that your training has oscillated in one direction to another following esoteric or unrelated topics to the next idea and forever moving forward in a non-linear fashion (Figure 3.1A). But that is the reality in creating a trajectory. That is real life. However, the impression you have to give when you are telling your career story, interviewing, or networking is that you have always been on a singular trajectory; a straight line where you started at one point, gained experiences that are relevant to a certain career, and you are going to continue on that path to get where you want to be (Figure 3.1B). Accordingly, the reality is that your experiences are all over the place; thus, the impression you want to give when talking about yourselves is that you are very focused and have always been on a certain trajectory.

FIGURE 3.1
(A) The "reality" of your career trajectory up to this point. (B) The "impression" of your career trajectory that you want to convey to others. Image provided by the author.

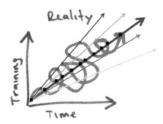

FIGURE 3.2
The data cloud and multiple trajectories that your experience can generate. Image provided by the author.

Changing Trajectory via Data Points

The take-home message for this book, if you get nothing else out of this, is that your experience and interests (in research) will take you down many paths. I encourage you to always be collecting that data, and there is no such thing as wasted effort when gaining experience. You have moved from this place to that, researched this topic and that, and followed this interest to where you are now. So, if you were to plot a graph of your training experiences over time, you would have a non-linear, overlapping, and possibly looping line that signifies your career up to this point. What you have actually created, however, is not so much a connected line but a data cloud that represents distinct experiences and skills gained.

When you talk to people, prepare application materials for specific job opportunities, or interview, you can tell your story using these many varied data points. Having so many data points allows you to frame (or reframe) those experiences and tell those stories in a compelling way that, say, "I have always been interested in this topic and I am on this trajectory, however, I do have all of these other relevant experiences." You can also pivot across these data points to create non-traditional and sometimes unexpected trajectories (Figure 3.2). You have generated a lot of data points in your PhD experience and will continue to do so in your postdoc training. It is up to you to assemble those data points into a straightforward compelling career or P-A-R story (further explained in Chapter 9).

Self-Assessment and Reflection

When leveraging your PhD, you need to be very active. Research progress, career advancement, and professional development are not passive processes. It involves some self-assessment, convincing yourself of your value, and persuading others that you are worth investing in and eventually hiring.

Assessing Your Situation

To know where you are going, you must also know where you have been and where you are now. You can do this by assessing your situation, your trajectory, your training, and yourself. This ensures that you are checking in with yourself and others. To ensure that you transition to a job (or career) that is a good fit, you want to be assessing yourself at regular intervals throughout your training. It is also helpful to do a self-assessment when you are faced with a big or very important decision to make. The three components of self-awareness and self-assessment that help to ensure career satisfaction are skills, interests, and values (Figure 3.3). Moving forward, you want to understand how these pillars intersect, synergize, and even diverge.

By knowing what you are good at, you can identify your skills. As outlined in Chapter 2, you should at the very least be building and honing the NPA Core Competencies within your postdoc toolkit. To review, these are research and problem-solving, scientific knowledge, responsible conduct of research, communication skills, management and leadership skills, professionalism, career advancement, and transferable soft skills. By finding what you enjoy doing, you can follow your interests and recognize themes throughout your work. This also reflects and informs your activities and affinities while keeping you engaged and satisfied. Following your interests provides focus for your ideas and bridges your diverse pursuits. Essentially, these are the things that get you out of bed in the morning, excited to seize the day.

By understanding what matters most, you can develop and solidify your core values. Once shaped, your values allow you to distill opportunities (or challenges) down to a few essential fundamental qualities. You can more easily prioritize what you require to be successful (and happy), what you are willing to put up with, what you can (and cannot) survive without, and what are ultimately your deal-breakers. When reflecting, you will be able to better assess how your values connect (or disconnect) with what you choose to participate in or pursue.

When you assess yourself in this way, you begin to analyze and put your activities, ideas, and motivations into perspective. This reflective process

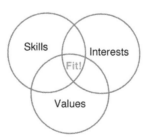

FIGURE 3.3
The convergence of your skills, interests, and values can lead to long-term career fit and ultimately satisfaction. Image provided by the author.

becomes an affirmation of your strengths and competencies. But, perhaps more importantly, you become a bit more aware of the gaps and areas of improvement in your training. It also prevents tunnel vision as you do not want to have blind spots, or be blindsided, around areas with which you should have specific expertise. You can avoid the embarrassment of not being able to fully answer questions about your science or yourself. In essence, when you do any kind of self-assessment, you are actually embarking on a journey to discover a very satisfying fit for your next job or career by taking into account the whole of your experience and not just a single dimension of it.

There may be times when you need to navigate a situation where an opportunity you want to pursue (or are currently performing) engages only one or two of the three self-awareness pillars and you are not particularly excited about doing the actual work. Furthermore, you might find it difficult to account for a missing pillar when seeking a good fit. This is where I remind you that your first job out of postdoc will probably not be your last and that you will likely have to progress along your career trajectory in a stepwise manner assuming you are moving each step closer to that satisfying fit of skills, interests, and values. Realistically, you need to have a job so you can be paid. The easiest first step is to leverage your most recent and active skills, because what you are good at can relatively easily lead to the next job. Once you have a job in the field or sector you are pursuing, you can use those skills to follow your interests. Then you can allow your skills and interests to dictate where you move to next, all the while aligning your career advancement to meet your values.

Developing Plans and Accomplishing Goals

Contemplating the Steps

Once you have done some self-assessment to understand what you need, you can now create a plan to move forward while addressing any relevant gaps in your training. By establishing achievable goals, tracking your progress, and celebrating your successes, you will not only successfully complete your aims but also have confidence in setting the next goal as you repeat the process. To me, there are several steps that help you get to the point of actually making plans and setting and attaining goals. You must focus on clarifying the steps needed to accomplish what you set out to do. You begin with the idea of doing something, the literal thinking of it. Next, you speak it into existence and make it real by committing it to paper. Sharing these plans and communicating your ideas to others will add accountability. By being your own advocate and using your network to help refine your endeavors you might uncover even more objectives to pursue and goals to set. Involving others will speed up the process by engaging them to consider your goals, uncover any challenges, and share solutions with you. You can execute your plan by taking advantage of the feedback you receive from your network. Then it is up to you to actually do the thing you planned to do. Remember, it takes practice to do this well along with time to convince yourself that you are doing something worthwhile.

Planning and Goalsetting

There is a broad spectrum of pithy opinions when it comes to planning and goalsetting, ranging from their complete futility to their practicality to their absolute essentiality. The oft-quoted Yiddish proverb gloomily reminds us that as "man plans, God laughs." According to Robert Burns in his poem *To a Mouse*, even "the best laid schemes of mice and men often go awry." The French poet and author Antoine de Saint-Exupéry warns, "a goal without a plan is just a wish." Whereas American author Zig Ziglar shares his optimism in "a goal properly set is halfway reached." One of the founding fathers of the US, Benjamin Franklin, demonstrates the necessity of planning by declaring, "by failing to prepare, you are preparing to fail." And finally, my ever-patient wife Heidi, with the incisive wisdom only a loving spouse can provide, told me as I shared my hope of completing this book, "If you do not have a goal, you will not accomplish it."

Still, I see a plan as a series of related goals, and the act of planning is just an advanced exercise in goalsetting. Many of you, and I include myself, may claim that you are not good at planning or goalsetting, but I contend that you are constantly making plans and setting goals nonetheless. I have realized, however, that it is not whether we are good or bad at it, but how *SMART* we are about the process. In using the SMART process, you are taking methodical steps to develop a **S**pecific, **M**easurable, **A**chievable, **R**elevant, and **T**ime-bound goal. The first action is to name it. A *specific* goal is focused and unambiguous. You should think about this in terms of a "minimum viable goal" and work from there. Next, you quantify it. A *measurable* goal inherently provides built-in benchmarks and milestones marking your progress until it is completed. You then need to be able to actually do it. An *achievable* goal is attainable given that realistic actions can be taken when considering difficulty, resources, and timeframe. You need to be sure you can apply it. A *relevant* goal is significant to you and others. You have to give the goal a meaning beyond merely accomplishing it. And finally, you have to actually finish it. A *time-bound* goal has an explicit deadline. You must create a distinct end point otherwise you might remain toiling for an inappropriate amount of time at the expense of more important priorities.

SMART AF Goals

In my work, I found that while the standard stages of the SMART goal process are invaluable, I needed to further bolster them by introducing **A**ccountable and **F**ortitudinous steps. Thus, making your goals *SMART AF*. When embarking on a goal, you need to check it. An *accountable* goal will help you stay on task. The most efficient way to do this is to make a concrete, specific plan, in essence, just following the SMART goal protocol intuitively includes accountability. You can further hold yourself to account by identifying the stakeholders that are interested in the outcome of your goal and share responsibility with them. Celebrating successes and milestones helps you sustain your enthusiasm for completing the goal while ensuring that you are strategically reviewing each step as well as revising the overall goal as you go. You will encounter many challenges along the way to finishing your goal, and you will therefore need to reinforce it. A *fortitudinous* goal has elements of courage and resilience embedded into it so that you can accomplish your goal while facing adversity. To maintain your steadfast commitment to completion, you need to fully understand the benefits of achieving your goal. By anticipating any potential obstacles and finding relative solutions, you are able to identify alternative means and methods for achievement. It is natural to think of success as the only option but acknowledging hardships and learning from your failures is essential for forging new key skills as well as revising goals for the better. However, you should know who your allies are and the people you can ask for help when you do get stuck. Just remember, take one step at a time and trust the process that you carefully planned out.

SMART AF Goals

Specific – focused and unambiguous
Measurable – milestones marking your progress
Achievable – realistic actions can be taken
Relevant – significant to you and others
Time-bound – an explicit deadline
Accountable – helps to stay on task
Fortitudinous – face adversity with courage

You will have many occasions to implement SMART AF goals. For example, a set of career advancement goals may include expanding your professional network, updating your CV/résumé, identifying new mentors and advisors, or seeking informational interviewing prospects. Additional cases for implementing SMART goals could be around the completion of some piece of your project. You could create a plan to finally finish that stubborn statistical analysis for your CRISPR experiments where you would present the results at a department meeting or include them in a conference proposal. You could make progress on your goal to draft, revise, and finally submit the paper that has been sitting on your to-do list. You could also develop a goalsetting process around expanding your presentation skills. You might first get foundational knowledge of presentation best practices by reading a book, talking to an expert, or taking a course. Then you would look for ways to practice and hone that new skill such as giving a short talk on your research or an unrelated topic. Finally, you might ask others to assess your level of mastery and critique your presentation skills.

Maintaining Progress

To sustain the progress you have made, you need to practice patience and maintain motivation. Give yourself the grace that you would give others on an unknown path to learning and building a reputation. And remember, it takes time and effort to do this. Subsequently, there are a couple of different ways to go about getting and staying motivated. As in your research, the earlier you seek, clarify, and verify any information, advice, or guidance you gain along your progress the better. Of course, you need to have a plan while determining the next thing that needs to be done. Try your best to keep things as simple as possible. Regardless of your approach, you have to realize it is going to take time for you to get good at anything. It is going to take time to finish your projects and revise your application materials. It is going to take time to learn how to interview well. It is going to take time to actually convince yourself that you are capable of transitioning. It is going to take time to appreciate your worth. And it is going to take time to find a job with a good fit. You do not have to go it alone while you gather your thoughts and collect your materials. You have people and resources, maybe even a career and professional development office like mine, at your own institution. If not, you can use resources as an alum from places you have matriculated. A thing I realized early was that if I have questions or issues with something in particular, there are countless others with those same or similar needs. Why not share your worries, reach out to others, and access resources as well? You can literally gather a community around you to help in your research as well as your job search. Maintaining progress and staying motivated in your transition is difficult at best and nigh impossible without the input and feedback of others. For example, I cannot tell you how many résumés, CVs, cover letters, research plans, and more that I have critiqued. I did not just help them in that moment and then send them away never to hear from them again. No. I insisted on having a conversation about the document or process that inevitably brought out more details, thus adding more context to the material. So, I return to advising you to be patient and give yourself room to grow. Just allowing yourself the necessary time to do this and to do it well should minimize stress while keeping you motivated to continue.

Getting a Job

Multiple Career Paths

Most trainees I counsel seem resigned to only two career tracks: *academia (i.e., teaching and research)* or *industry (everything else)*. While an academic career provides a known landscape, compatible training, and an intellectually rich environment, it is also highly competitive, growth-limited, and grants-driven. If this path does not pan out, the default alternative is industry with its potential for growth in a fast-paced and highly collaborative environment. However, industry is also an unfamiliar landscape that is still highly competitive, somewhat unpredictable, and business-driven. Obviously, the post-PhD scientific workforce is much more nuanced than described above, therefore, I argue that the trick to finding your endpoint is for you to recognize that there are more than two career trajectories. By considering all of your options, you can take control of your training and begin to make smarter, more focused decisions regarding research directions and professional development.

At this point, you should be considering all of your career options, some of which you may not yet have realized as possible. For example, I shared earlier in the book that my current role simply did not exist until 2007. At that time, I was exploring my own career options and transitioning from a PhD candidate to postdoc fellow. While it sounds trite to say that you can do just about anything you want as a PhD, it does not change the fact that it is true. Having said that, sometimes, eliminating options can be as helpful as adding career options to explore. I am unable to enumerate all of those career possibilities, but I do share some major areas of potential career pursuits in a few paragraphs.

First, I want to dispel the myth that you are only qualified to do academic research and/or teaching. If that is indeed what you want to do, you are well-positioned already. Having worked with graduate students and postdocs, as well as from my own experience, I fully understand both the attraction and repulsion of chasing a career in academia, especially as research or teaching faculty. Regardless of where you end up in your career, academia is the place you started. It is where you pursued higher education, received a PhD, and where the vast majority of postdoc training takes place. Having taken the only path available to you so far, most of you have merely witnessed teaching, science, and research being done in academic settings and thus only really know a very narrow characterization for career outcomes. There is no doubt that the academic research enterprise provides an intellectually rich environment with a familiar landscape and likeminded training. However, we also know that this same environment can be highly competitive to the point of toxicity and, in most cases, is the historical product of a white, male-dominated enterprise. This is not to mention the recent trend of limited faculty openings and shrinking funding opportunities in that last two decades. While strides are being made in diversity, equity, inclusion, and belonging, it comes as no surprise that many PhDs and postdocs choose to pursue careers outside the academy.

Beyond the "traditional" academic route, you have many options across higher ed, the public and private sectors, as well as industry. I have seen many postdocs go on to pursue higher ed leadership, biotech, pharma, entrepreneurship, product development, marketing, business development, consulting, venture capital, policy, outreach, and non-profit. The list goes on and on because the opportunities that your PhD and postdoc research training set you up for are more than just academic research or even industry. To continue, I have helped postdocs pursue careers in science writing, journalism, editorial, publishing, regulatory science, intellectual property (IP), patent law, tech transfer, K-12+ teaching, curriculum development, and government and clinical research. You can even follow my career path into student, postdoc, or faculty affairs as well as my future path (fingers crossed) in executive-level administration.

PhDs and Postdocs have many career options to consider.

Academia: Academic/Clinical Research; Undergraduate/Higher Ed Teaching; Executive/Academic Administration; Student/Postdoc/Faculty Affairs; K-12+ Teaching; Curriculum Development

Industry: Biotech/Pharma Research; Product Development/Marketing; Entrepreneurship/Business Development; Consulting; Venture Capital

Government: Government Research; Science Policy; Science Outreach & Education; Non-Profit

Communication: Science Writing; Journalism; Editorial; Publishing

Law: Regulatory Science; Intellectual Property; Patent Law; Technology Development/Transfer

As I said, not all careers are represented above, there are dozens if not hundreds of other types of jobs you can do as a PhD. While it is not the scope of this book, I will offer a short list of resources at the end that many of my advisees have utilized with success to explore diverse career pathways.

Part 2

The Middle

4
Situational Awareness

Postdoc Pain/Pivot Points

Postdoc training is fraught with pain points, stressors, and pitfalls. This is not meant to be gloom and doom, but to ensure that you have your eyes wide open to the realities of the academic research enterprise. Most of the conflicts you will face as a postdoc will not come from doing your research but will be a result of difficult relationships and misunderstood feelings (yours and others). You need to make sure you are identifying these situations where you may feel like you are out of control or in conflict. The idea is that when you identify professional pain points, such as an inactive network, an unknown career target, or uncertain marketable skills, you try to pivot out of these situations instead of spiraling even further. What you need to do is recognize that you are in a difficult situation and begin to shift from that mindset and into a more positive space.

> *Professional Pain Points* – Inactive network. Unknown career target. Unpolished career story. Unsure of marketable skills.
> *Professional Pivot Points* – Active network. Clarified career target(s). Shared career story. Honed marketable skills.

As for personal pain points, I see a lot of postdocs coming into my office with analysis paralysis, where they are faced with too many choices or they are faced with a choice that paralyzes them because there are so many implications to making that choice. There may be unrealistic expectations either placed on you or unrealistic expectations that you have for yourself in your research, job search, or networking. You need to check with yourself and examine whether this is truly what you want. You must be careful on your postdoc journey because you could just get tired, exhausted, or worse, burned out. This is a real problem that may lead you to consider transitioning out of academia, away from the bench, or out of science altogether. In fact, I have actually had postdocs in difficult but seemingly manageable situations come to me in full crisis mode because they were so weary and so narrowly focused that they could not see a satisfying solution. And this was mostly due to the fact that it had been months or even years since they had any significant time off. Again, gathering your wits, you might just realize all you really need in that situation is a break or time off to gain perspective and recharge.

> *Personal Pain Points* – Analysis paralysis. Unrealistic expectations. Exhausted, jaded, or burnt out.
> *Personal Pivot Points* – Evidence-based decision-making. Set realistic expectations. Built resilience.

Other pain points you want to realize are in your project. You might be near your appointment or project end date or your overly complex project keeps expanding in scope. You might also feel that you have incomplete accomplishments. Unfortunately, it is common to experience uneven or neglectful mentor engagement. If you have a perception that you are not getting what you need out of your mentor-mentee relationship, you need to begin to work on that and realize that you should begin to "manage up" to get what you want from your faculty mentor. Again, these are just pain points where you need to acknowledge that this might be a potentially negative situation, recognize it, and take action to pivot into a more positive space.

> *Project Pain Points* – Scope and project creep. Incomplete achievements and accomplishments. Little or no mentor engagement.
> *Project Pivot Points* – Established parameters, deadlines, deliverables. Completed achievements. Increased and multiple mentor engagement

The Tough Postdoc Environment

Dr. Ebony McGee, in her book *Black, Brown, and Bruised: How Racialized STEM Education Stifles Innovation,* succinctly describes what it can feel like in STEM training, "I know I have to work twice as hard and hope that makes me good enough ... " Likewise, Dr. McGee's quote resonates with me as I view it through the lens of postdoc research training since the postdoc environment can make you feel like an inadequate imposter; like you are a phony; like you have to constantly prove your worth; like you have to out work everyone around you just to stay even. While her book focuses on the educational challenges of historically underrepresented groups (URGs) in STEM, especially those of color, similar attributes can be observed in how you, as a postdoc, are treated within the academic hierarchy. Thus, for some, instead of experiencing a pleasant stress-free postdoctoral fellowship, you encounter prolonged internal and external academic research pressure where you expend higher levels of effort and energy at the potential cost of your physical, mental, and financial health. Furthermore, any person pursuing a postdoc that is part of a URG or from another country is likely disproportionately impacted. This may sound hyperbolic, but it is a very real danger. Postdocs, due to your varying levels of support within the lab, department, and institution, are at risk of being unsupported, overworked, or outright exploited. The recent COVID-19 pandemic continues to exacerbate and highlight these disparities.

On top of this, you are also in danger of experiencing harassment, bullying, and discrimination, with your faculty mentors being the most likely perpetrators. As a postdoc, you will often be unsure of the resources available to address the above concerns because of your sometimes ambiguous status, eligibility, or access. Even when appropriate resources are offered, many of you will remain reluctant to seek help due to any number of reasons, but most deal with skewed power dynamics and fear of potential retribution. This is why it is vitally important for you to fully vet your current or future mentors so as to promote safe and inclusive training environments where you feel like you belong and are supported to be your authentic self. It is especially poignant for international postdocs since you are uniquely beholden to the lab and institution due to your Visa sponsorship situation. Again, I write this not to scare you but to make you aware of, and prepare you for, situations you might find yourself in currently or in the future. The second half of this chapter outlines some strategies you can use to address and allay those fears, while Chapter 5 provides in-depth approaches to managing your mentor-mentee relationships. Regardless, there should be considerable resources within your institution to assist you in dealing with difficult environments and people including but not limited to the Ombuds, human resources, international office, department leadership, the postdoc office or association, or the research and professional integrity office.

Potential Resources for Conflict, Harassment, and Discrimination Resolution

Postdoc Office or Association – Usually the first stop of information-gathering, these groups work to enhance the experience of postdoctoral fellows while promoting a sense of connectedness throughout the postdoctoral research community. They also aim to support the health and wellbeing, research advancement, and professional development of postdocs.

Ombuds Office – As an impartial conflict resolution office, the Ombuds strives for the fair and equitable treatment of people in the midst of a crisis. The Ombuds is usually independent of any existing administrative or academic structures and is confidential, neutral and, as such, does not advocate for any one individual or point of view.

International Office – Usually a centralized administrative office that offers services to foreign students and scholars across the institution. It provides information on a wide range of topics, including visas, work permits, travel, financial questions, social and cultural differences, and personal concerns.

Human Resources and Employee Development and Wellness Office – These offices usually provide information, resources, referrals, educational programs, and support for employees in balancing professional lives with personal concerns.

Employee Assistance Program (EAP) – Usually a contracted wellness provider that can connect faculty, staff, and postdocs with a rich network of resources that can help employees manage the competing demands of work and life.

Title IX, Gender Equity, and Diversity Offices – Many institutions have designated offices or personnel that handle policies and procedures designed to provide prompt and equitable support for or investigation into concerns regarding harassment, bullying, and misconduct.

Research and Professional Integrity Office – These offices promote best practices and standards of professional conduct and support grievance processes in accordance with institutional professional integrity policies. The office is also usually responsible for investigating concerns related to research integrity as well as fostering a safe and healthy research environment.

Postdoc Needs and Worries

You will also have to reconcile with the needs and worries that arise as you progress through your training from PhD to postdoc and beyond to your early career. You will ask yourself over and over how can I make my research and science better? What will get me published? How do I stay funded? What will get me a job or a promotion? As a PhD student you probably worried about becoming overeducated, being exploited, wasting time, and losing money while you sought much-needed attentive mentoring and a path to publication. In your research training, you needed to learn methods, experimental design, and data interpretation. In your professional training, you needed to practice your communication and writing skills. You also needed options in your potential exit points and career exploration.

As a postdoc, you will continue to fear and seek out much of what you did as a grad student, both on broader and deeper levels. At this point in your training, you are probably feeling anxious about exploitation, lack of independence, overtraining and underemployment, low salary, caring for your family, and ambiguity around your career readiness. You will still need effective mentorship and progress toward publications but now you really require a clear path to independence. Beyond just building research awareness, you are now moving into scientific mastery of methods, statistical analysis, and experimental reproducibility. Your postdoc training now includes professional development and a notion of your job search as you begin to explore career outcomes and exit points. You are now seeking necessary leadership, mentoring, and management experience as you continue to grow your network, find funding, and advocate for your own research and career vision.

Within the first several years beyond your postdoc, you will have to continue to do everything mentioned so far in your early careers. In addition, you need to cultivate your own career advancement, growth, and promotion all the while creating harmony within your life and work. You are also going to be much more anxious about attaining sustained funding, being paid what you are worth, and continuing to grow and provide for your loved ones.

Stopping the Negativity Spiral

Fear and anxiety tend to rear their ugly heads especially when facing hard decisions, conflicts, hardships, or transitions. Those situations may cause deep distress and/or pervasive negative self-imagery for some of you. When this happens, you must recognize what is happening and stop the negativity spiral as quickly as possible so you can pivot to a place of objectivity or even optimism. You can start by processing your stress and taking steps to move beyond the cause of your anguish. Try not to re-live or re-litigate the negative situation, but look at these circumstances as growth opportunities, if possible. The lessons you learn through difficult times sharpen your resolve and broaden your perspective. You begin to think strategically about the next steps, from cost-benefit analyses to SMART AF goalsetting. After assessment and reflection, you can now re-contextualize your situation by telling neutral-to-positive stories about moving *forward*, instead of *escaping* from something.

There is a relatively loose sequence you can follow when facing a difficult or complicated situation. Definitely do not panic, realize you are not alone, and use your resources. You will discover that good, thoughtful, and well-reasoned decisions do not come from a place of fear. Recognizing that you do not know everything, you should engage in your research training (i.e., the scientific method) where one of the first things you do is gather information and data. You should reconnect with your network and seek advice from trusted colleagues and former mentors. By activating your existing network, you can let them know what is happening and what you plan to do next. You will have to consider the timeline in which you are operating and work backward from the endpoint.

Your postdoc experience, as well as that of your PhD, can be isolating at times, but you must not try to do everything by yourself. Most institutions have layers of support, and I encourage you to leverage all the resources at hand. After information-gathering, you will need to process and verify what you have learned by reaching out again for help, this time by talking to career coaches and other advisors who may be able to give unbiased perspectives on your specific situation. This is also an opportunity for self-reflection and self-improvement where, as outlined in Chapter 3, you can inventory your credentials, skills, experience, and interests. There may be space in your timeline to address real or perceived gaps in your training and education. If doing so, I advise you to pursue multipurpose, cross-training opportunities that offer versatile skills development.

At this point, the next steps of the sequence, and the advice that follows, diverge depending on the urgency of your situation as well as its fixability. It is possible that your initial negative response ends up being a low-stakes overreaction that, once acknowledged, can be quickly resolved. A clash could also bring to light a pervasive problem and actually leads to a satisfactory overhaul of the situation, salvaging of a relationship, and continuing growth. An alternative reaction may cause you to fast-track your career or job transition that, depending on the state of the relationship with your faculty mentor, can either be encouraged or not supported at all. Unfortunately, one of the more dire consequences of poor conditions or a falling out may even result in your speedy departure (or outright removal) from the situation, leaving you to quickly find a different lab, faculty mentor, postdoc fellowship, or job.

Fear of "Throwing Away" Your PhD

As outlined above, you are going to be under enormous pressure, both formative and destructive. If you choose to leave academic research for the reasons listed above and/or for another path altogether, you may be conflicted about or have others question your decision. I shared earlier in the book that I had a similar experience. I spent years fighting my fears and following my interests toward my career path alongside performing my research. All that time I was suppressing my own doubts while convincing my wife and family that I had not wasted ten years of my life and theirs in pursuing PhD and postdoc training. It took me a while to accept the fact that once you receive your PhD, you will always have a PhD, and no one is out there trying to take it away.

The sooner you realize that there is no longer an expected or default PhD career trajectory, the sooner you can reconcile the warring pessimism of throwing away your scientific career with the optimism of pursuing a satisfactory one. As I said, they are not going to revoke your PhD nor can anyone take away your extensive scientific training and research experience. For example, I am no longer doing research but I am still using the skills I honed as a graduate student and a postdoc to solve problems. I earned a PhD because I solved a very specific problem no one else knew about (or maybe even cared about). Further, I have been able to play an active role in academia by advocating for postdoctoral fellows, working to improve their training environment, and advising them into diverse independent career pathways. But I digress. The PhD and postdoc process taught me how to address challenges in a systematic way and I am still doing what I was trained to do, but now I am just doing it in a (very) different setting. I have not wasted anything, and neither have you. You are always going to have the core values of the scientific method and the years you spent in graduate and postdoc training.

Normalizing Struggle and Failure

Having come to a decision, you may feel that you made a grave mistake. You may even worry that once you move away from the warm cocoon of academic training it is only going to get more difficult and that you might ultimately fail. To that I say, you will be okay since you have also been trained by research to deal with failure. In fact, as a scientist, your whole existence is to prove the null hypothesis, to actually fail and demonstrate that anything you do to perturb a system will have zero effect. In that sense, you have been "failing" in a laboratory for many years. Many of your experiments have not worked, for any number of reasons, and you still found a way to move your project forward toward success. You are familiar with failure at this point. You have been in a dark box of struggle, ambiguity, and failure and you are constantly learning to get out of that dark box or more significantly, move that dark box into the light. The trick is to remain curious as you change your perspective to understand that the subject of your experimentation is now yourself and your career (rather than mice, cells, or nucleic acids).

To continue to normalize failure, you must grasp that everyone around you is struggling in some way or another. Most of the time you only see or hear about the successes of your peers because very few people actually disclose their challenges and hardships. For example, you may be familiar with the concept of a "CV of Failures" popularized by Dr. Johannes Haushofer where he quite humorously and honestly shared all of the rejections he received from PhD programs, grants, papers, and job applications. He succinctly demonstrated that most every success is preceded by several failed attempts. Thus, admitting your limitations, accepting your failures, and sharing your struggles can be important steps in advancing toward success. In addition, having a growth mindset where all experiences, good and bad in life or lab, are learning opportunities that build self-awareness, minimize blind spots, expand boundaries, and foster resilience.

Imposter Syndrome: Managing Your Inner Dialogue

As mentioned earlier, the scientific research enterprise can make you question your competence and your value. It can make you feel like you do not belong, like you are going to be found out as an imposter. Worse, academic research fosters intense competition often at the cost of comradery. Due to this, you must resist the natural urge to compare yourself, your productivity, your progress, or your predicament to that of your peers and senior colleagues. This is a recipe for disappointment with no scenario that ends well for you. Not only can you never fully know their circumstances; you are very likely making false assumptions about their thoughts, feelings, and relative successes. Doing so exaggerates their contributions while also downplaying your own. This may lead you to further distort your reality and heighten your imposter syndrome.

You can turn this on its head with some effort. Just like you can be your own harshest critic, you can be your own greatest advocate. First, you must engage your *inner critic* by tracking your self-dialogue, challenging your negative assertions, and inviting that same critic to be part of the learning process. It is then essential to begin changing the story you tell yourself. You can start by writing out the current narrative that is running through your head. Make sure you list the words and phrases that evoke a reaction. Take a moment to check the veracity of these words as they describe you and your actions. Then imagine what a supportive story would tell about you by replacing the untrue and negative words. Now compose a new and accurate story to tell yourself. However, I would go a step further than just a correct story, I suggest that you craft an inspiring story to tell yourself. One where you are not a victim or villain, but the hero. This story will eventually form the larger narrative of your career that you share aloud with others.

When re-drafting your inner dialogue, take advantage of the fact that hindsight is 20/20 as you can now reflect on the past and place its true effect on your present circumstance and future options in context. Your internal (and eventual external) discourse should include several elements of creative hero-origin storytelling. It should be emotive, optimistic, future-focused, confident, and, of course, inspiring. Your self-story should be full of concrete details as you remind yourself of your unique interests, goals, and motivations. While this may sound cliché, being the hero of your own stories is nothing but a distillation of your own authentic resilience through failures, hardships, challenges, and barriers. However, be sure not to inflate your story, show a lack of humility, or stray into toxic positivity. Regardless, the story you tell yourself should specifically demonstrate a growth mindset where you continue to learn lessons from your adventures. Below, I share examples of self-dialogue that show how you can exchange your narrative of negative self-imagery by composing an accurate and inspiring story to even optimistic storytelling for the future. When others hear your new story of perseverance, passion, and progress, they will be compelled to be your advocate.

> *Negative self-imagery:* I am a poor postdoc without retirement benefits, misunderstood by the world, fighting uphill against obstacles, with a PI who hates me. I am all alone and have few career options, struggling for leftovers.
>
> *Accurate/inspiring story:* I am a thoughtful and hardworking postdoc with an intellectually challenging project that has potential for great impact. While this period in my training is difficult, I have the tools, motivation, and resources to see it through.
>
> *Optimistic future-focused self-advocacy:* I will be a successful project management professional with incredible benefits, independence, and overcoming challenges with a supervisor who supports me. I will be at the center of a giant network with opportunities for growth and advancement.

Resilience and Mental Wellness

In truth, difficult situations are as much a furnace for growth as they are for burnout. The postdoc environment can be a fiery crucible of hardship, ambiguity, and conflict, but it can also forge self-awareness, resilience, and new perspectives. There is strength to be found in your community since postdocs are a diverse, high-achieving, and globally mobile group. That global reach drives the scientific enterprise and its innovation while creating opportunities for progress as well as friction. The key to surviving and eventually thriving as a postdoc is boosting your resilience and normalizing failure. You are expected to find ways to take care of yourself by maintaining your physical fitness and mental health as you continue to establish reasonable work/life integration. You have to pursue activities, make connections, and nurture relationships outside of your postdoc environment. According to the American Psychological Association, in building resilience as well as bolstering mental health and wellness you begin to establish a positive yet realistic perspective of life and lab. Instead of viewing crises as insurmountable, you can now see their true form as problems awaiting to be solved. As you accept that the concept of change and conflict are not always malevolent forces, you allow yourself to take decisive actions while looking for further opportunities for self-discovery. This becomes a positive feedback loop, a virtuous cycle of self-care, where you nurture a positive inner dialogue, keep things in perspective, and maintain a hopeful outlook. Additionally, once you have gained some semblance of wellbeing, you may then be able to look out for and support others, possibly even intervening on their behalf when they are facing hardship.

As I have shared many times, I was able to overcome mistakes and challenges by developing survival skills and resilience. I have learned to proceed with empathy and kindness while assuming the best intentions of others. I just kept showing up and continued to do the work. I also recognized my limitations and built a team of peers, colleagues, and collaborators around me. In doing so, I cultivated mentors and allies as I also improved myself and relationships. I maintained, and even increased contact, in some cases, with friends and family. Yet as I followed my strengths and interests, I was able to learn from my mistakes and seek out silver linings.

For example, during the COVID shutdown and subsequent protracted return to normalcy, we all faced personal and professional hardships. In this time, I was able to appreciate some silver linings without gaslighting myself with toxic positivity. Between March 10 and 11, 2020, my office literally went remote overnight. After grasping the gravity of the situation and understanding that we were not going to return anytime soon, I realized that I now had an extra three hours a day since I no longer had to commute. Regardless, my family's dynamic and routine had been massively disrupted, and there was little to no personal time or space since we were all home together seven days a week. Nevertheless, I was now able to more equitably help out around the house and I began cooking for the first time. We also leaned into our family's shared interests and developed richer and more respectful relationships. At work, it was a time of financial insecurity for the school, and my office's operating budget was slashed. However, this forced us to work more closely with other offices, and we were able to shift to collaborative and co-sponsored programming across institutions. Because of COVID, there was a new kind of fear, uncertainty, and existential dread. A silver lining I found in this instance was that irrespective of how bad it may be, it was at least a shared experience everyone had and I was not alone. Similarly, COVID laid bare many social, medical, and historical inequalities and the disparity in postdoc support came to light. While I had been aware of and working on these for years, my office, myself, and postdocs now had a seat at the table and a voice in the room when decisions were being made.

The message I hope you take away from all of this is that no matter what you may be facing, keep an open mind, be open to possibilities, and stay in the moment. If you are able to learn from your misadventures, you can make impactful contributions while exceeding all expectations. And always remember that you are not alone, that you can count on friends, colleagues, and mentors to be there in good times and bad.

5
Navigating through Your Postdoc

Establishing Ground Rules and Expectations

Every lab will have its own norms and work expectations for you as a trainee. At a minimum, you must dedicate the time and effort required to make continued progress on your research while actively contributing to the lab through mutual sharing, mentoring, preparation, and sincere effort. It is expected that you will gain the knowledge and experience necessary to advance your career through publications, independent investigation, and development of your network. If your faculty mentor believes that there are obstacles preventing you from fulfilling this commitment, they should discuss it with you and/or refer you to appropriate resources for advice and further guidance. While it may be necessary to be in the lab at any given time, day or night, you should never be expected to work seven days a week for extended periods and must be allowed time for vacations, holidays, and other breaks from lab work. Since research hours are usually not recorded, you should work with your faculty mentor to manage your time, experimental schedule, professional development, and research progress. If at all possible, required activities such as lab meetings should be held during regular business hours to accommodate any family and other important outside obligations you and others may have.

Your Expectations

As you may recall from Chapter 2, there are many things you must consider when selecting a mentor and research environment. In addition to aligning research interests and scientific compatibility, you must also establish ground rules and expectations for your subsequent training and completion thereof. You should ask what their definition(s) of success is and where their successful (and unsuccessful) former trainees have continued their respective careers. You should not only clarify your faculty mentor's long-term expectations for you in your training but also in the near- and mid-terms, always with an emphasis on providing a path to independence. You should strive for complementary mentoring and management styles while inquiring about how projects are allocated as well as the financial stability of the position. You should know whether you need to write grants to secure funding as well as if you have the requisite access to proper facilities and instrumentation to fully pursue your project. The lab should have a widespread recent track record of publishing, and it should be apparent what the manuscript writing and submission process is. Importantly, you should expect to join a well-adjusted and supportive training environment with good-to-high morale. That environment should be grounded in mutual respect and driven by psychological safety and intellectual curiosity. You deserve no less than genuine mentor engagement, freedom to fail, and an opportunity to grow.

Their Expectations

Your mentors also have many expectations for you. Not all mentors are created equal though, and some may not explicitly communicate their expectations hoping that you either learn on the job, from watching others, or by osmosis. Simply put, your mentors expect you to do your research and take ownership of your project(s). They want you to be resourceful, organized, and generous with your time, materials, and advice. You should seek to be a good lab citizen by mentoring lab mates, exhibiting a collaborative spirit, and getting along with others. You are expected to strive for excellence while always giving your best effort. Your mentor wants you to be open to feedback and critical suggestions. These expectations, whether shared or implied, can be boiled down to the three overarching principles: be responsible, be respectful, and be industrious. Finally, your mentor does not expect to have to act as your caretaker so, as mentioned in Chapter 4, you are responsible for taking care of yourself, pursuing interests away from research, and nurturing relationships, as well as maintaining your physical and mental health.

Meeting Expectations

You and your faculty mentor(s) must have regularly scheduled meetings that not only cover project and experimental progress but also interpersonal, career, and professional development as well. This is how you clarify research responsibilities and understand how success therein is defined. This is also where perceived skill or experience gaps should be addressed and support offered to learn necessary skills while accomplishing your work.

In addition, an individualized development plan (IDP, discussed in the Mentorship and Individual Development Plans section) will guide you in your research, career exploration, and professional development along with identifying other relevant mentors whom you might be seeking out. You should be encouraged to discuss and explore desired career trajectories while also pursuing relevant training experiences throughout your postdoc tenure.

Mentorship and Individual Development Plans (IDPs)

Your postdoctoral training goals and expectations should be aligned with those of your faculty mentors. From the very start of the postdoc appointment, I strongly encourage training plan discussions between you and your faculty mentor(s). It is worth mentioning, however, that most PIs have not received instruction on how to manage or mentor effectively. Depending on where they are in their career, your faculty mentor may still be adapting their approach and learning as they go. Additionally, they face tremendous pressure working to keep the lab funded and productive. Please remember that they are human and have flaws as well. Because of this, there are templates, guidance, and training available for managing mentoring relationships within career and training offices as well as NIH, NSF, and other national scientific societies.

Since you are expected to engage in independent research while also developing skills and experiences that support your specific career goals and professional development, you should not do this alone without guidance. Having guidelines in place, whether formal or informal, will help define expectations for your mentor-mentee relationship, laboratory norms, research and career support, and professionalism. In combination with the IDP tool, these mentoring instruments can be designed to link your research goals with your career development and progress toward independence. Mentoring guidelines and IDPs are meant to foster an ongoing and recurring discussion that involves evaluation, goal setting, and feedback, with substantial input from both you and your faculty mentor. While the NIH and NSF require the use of IDPs for graduate students and postdoctoral researchers supported by specific training awards, I very fiercely recommend its implementation for all trainees, regardless of funding source. Furthermore, the uniform implementation of these bi-directional mentoring tools should be common practice, one that you may have to initiate.

Regardless, the core responsibility of your mentor is to oversee and guide your scientific and career progress. When taking you on as a postdoc, your advisor accepts responsibility for this unique role with respect to your scientific development. Mentors and trainees are expected to meet regularly, at a frequency that is appropriate for the ever-evolving stages of training. This is a joint effort between you and your faculty mentors, as you both share responsibility for creating an atmosphere that supports the many dimensions of one another's success.

The most useful IDP templates provide an open framework that can facilitate regular big-picture strategy sessions between you and your faculty mentor that addresses research and professional progress (Figure 5.1). This document should be filled out by both parties beforehand and discussed during the meeting, with the two of you eventually coming to a consensus

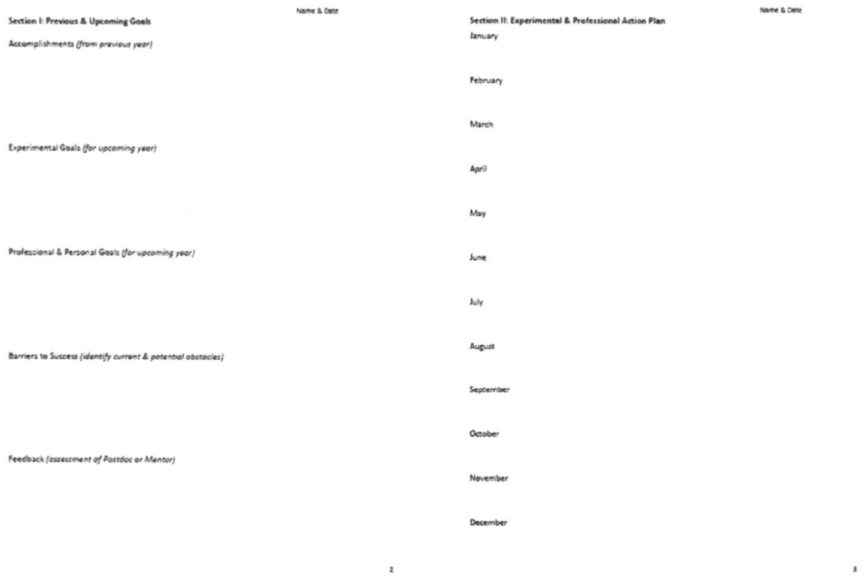
Section I: Previous & Upcoming Goals
Name & Date
Accomplishments *(from previous year)*
Experimental Goals *(for upcoming year)*
Professional & Personal Goals *(for upcoming year)*
Barriers to Success *(identify current & potential obstacles)*
Feedback *(assessment of Postdoc or Mentor)*
2

Section II: Experimental & Professional Action Plan
Name & Date
January
February
March
April
May
June
July
August
September
October
November
December
3

FIGURE 5.1
An IDP template. This example is formatted for maximal flexibility and can be edited to suit your specific needs. Image provided by the author.

on past accomplishments, future goals, and next steps. For instance, I can imagine a scenario where you have identified an opportunity that will allow you to gain and hone an important technique or skill and you feel that the two of you are not yet in agreement regarding it. A useful approach for you to take is aligning common interests by suggesting that by doing "this thing" or going to "that workshop" on presentations or grant writing, is going to make you a better and more productive researcher, for example. Adding that whatever you are learning is not going to take away from your project, timeline, or effort. In fact, it will actually enhance these things and potentially add further value for the entire lab. Then you share the actual steps you plan to take to accomplish such a thing.

A typical IDP outlines major topics of discussion, benchmarks for advancement, and identifies real or potential barriers to success along your training path. The IDP should also allow for an element of feedback where your faculty mentor can evaluate your performance and progress while at the same time letting you share issues related to research, training, or mentoring. You can make the IDP even more practical by incorporating a calendar that outlines and organizes your research and career development goals across the upcoming year. Briefly, the IDP meeting plays out in several steps. First, you and your faculty mentor should complete the IDP form separately and ahead of time. It is especially helpful if you provide this form with an updated CV for your mentor in advance of scheduling the conversation. The two of you should then meet and discuss your respective filled-in IDPs, making sure to review accomplishments, goals, barriers, and feedback. As the conversation comes to a close, you should have an agreed-upon framework toward making progress and meeting stated goals and objectives for the future. Finally, using the SMART AF tools discussed in Chapter 3, you can now implement your action plan, review your progress, revise as needed, and repeat the process next year.

Mentoring Up and Self-Advocacy

One of the most effective tips for successful mentoring relationships, and science in general, is to quit taking it personally (Q-TIP). Constructive criticism and failure are part of the research enterprise; use them to inform your next steps. Even though it may feel like it, your value is not determined by your research output. While aspects of research are very individual and you have both a personal and professional stake in the outcome(s), you need to realize that you can receive feedback on your work without taking it as a personal attack. That said, after a difficult, contentious, or confusing interaction, get a reality check. Your perception of a situation may not match reality, check in with a colleague before a further misunderstanding occurs. Equally, your understanding may be accurate but you may still need outside confirmation. It is helpful to get another perspective when things become uncertain or intense.

As stated above and before, you have to set goals and expectations together and re-evaluate often. It is important to set achievable goals and long-term expectations, just make sure to revisit them going forward. Effective communication is essential for dynamic mentoring relationships. This is predicated on a transparent exchange of ideas and expectations. With a full understanding of expectations, flexible SMART AF goals can be set, revised if needed, and achieved. To ensure this happens, it is your responsibility to always be prepared for a meeting and have all necessary meeting materials, sent ahead of time if possible. Preparing for a regularly scheduled meeting takes thought and you can be strategic by creating an agenda and following up with a summary of things discussed. Additionally, having organic, low-stakes, informal conversations are important opportunities for you and your mentor to check in with each other. However, not everyone is at their best in these situations. If your faculty mentor wants an impromptu meeting, ask for a few minutes to collect your thoughts and materials before sitting down with them, in private if necessary.

You are encouraged to ask for guidance but make sure you also have several ideas on hand. Expectations are such that you will not know everything about a project but when seeking input, but you must also contribute. As you grow in your expertise and independence, when asking for help or feedback, it is important that you explore potential solutions beforehand. Seek out perspectives from secondary mentors in addition to your primary faculty mentor. Cultivating multiple mentors is a common practice that allows broader discussion and support. These can range from informal non-hierarchical connections that give advice on everyday academic struggles to more formal faculty mentor-mentee relationships that offer substantial intellectual contributions to your project or career. By doing this, you also expand the potential pool of references for when you launch your job search.

Stay patient, mindful, and connected throughout your research experience as it can feel exasperating, tedious, and isolating at times. When this happens, it is important to remind yourself that struggles lead to learning. Building and maintaining a community, including your lab mates, your department, affinity groups, as well as your loved ones, will provide support. As in most supportive relationships, there is a give and take, an ebb and flow. Realize that mentoring is a two-way street. The mentoring relationship is an alliance built on bi-directional respect, open-mindedness, communication, and adaptation. You cannot be passive; you are expected to play an active role in managing the relationship. Be honest about your preferred style while also realizing they have their own way of doing things. The object is to align your mutual goals and be flexible as they evolve.

Remember that you are an adult and this is your career. You are ultimately responsible for your actions along with how your research and career progress. While you may be supported by your mentor and the institution, you are the ultimate stakeholder in your success. Therefore, you should advocate for yourself at every opportunity. When seeking advice and guidance, it is up to you whether to heed it or follow your own counsel. Use the resources available to you. Please be assured that you are not alone nor are you expected to know everything. Your institution has created an entire infrastructure to support you and your work. Human resources, the Ombuds, the postdoc office, department administration, and career services may be just a few of the resources you have to help you in your tenure as a postdoc.

Professional Development

Postdoc Skills and Competencies

You also want to evaluate the whole of your training. The National Postdoc Association created a list of postdoc core competencies that I shared in Chapter 2 and reiterate here. If you are in PhD or postdoc training, this is really what you should be striving to learn. They are basically broken out into your research skills and transferable skills. Research skills are knowledge of discipline, lab and experimental skills, and responsible conduct of research (RCR). Then, transferable skills are communication, professionalism, leadership, and management. If you are not working on those six major categories, at any one time in your training, you need to make sure that happens.

When you are considering what you need to be successful, you have to understand the skills that you and your potential future employers most desire. Many of you are very well versed on only one end of the skills continuum that contains hard and operational research skills such as methods, technology, quantitative, computational, experimental design, and data interpretation. However, what you need to really pay attention to is the other end of the spectrum and make sure that you are working all across this continuum of skills. Be sure you pay attention to your soft skills such as management, leadership, communication, and teamwork because those are the attributes that will set you apart from others that you may be competing against for jobs and advancement. Almost every other postdoc you encounter on the job market and in the next steps of your career will have basically mastered the operational and hard skill sets but may not have honed their soft skills. If you are doing well in those areas, you are already several steps ahead of your competition.

When you are looking for additional training inside and outside the lab, you want to make sure you are pursuing relevant opportunities that will give you the most bang for your buck across multiple potential career paths. What you do not want to do is become so narrowly focused on one thing that your training is no longer transferable. Chapter 6 provides an in-depth approach to recognizing and building transferable skills. I encourage you to think strategically about how you can diversify your skillsets inside and outside the lab, both within the confines of your research and beyond it. Consider what kind of specific activities you can pursue to develop these skills, whether you are mentoring students, building collaborations, giving presentations, leading a journal club, or attending a workshop offered by the postdoc office. When addressing these considerations, you must balance the time it takes to develop yourself with the time it takes to do research.

Build New Skills

Alongside honing existing skills, you should be looking for opportunities to fill gaps and add new skills. If you are interested in bolstering your teaching credentials, reach out to local colleges and universities to ask if you can observe a lesson or teach a class. I know for a certainty that faculty and department chairs at undergraduate institutions would welcome the chance to have you in their classrooms. You may even consider connecting with teachers and educators at K-12 schools, science centers, and even local libraries. If you are expanding your research portfolio, take a lab management course or learn a new cutting-edge technique that is potentially valuable now and in the future. You can attend a grant writing workshop, finish that data analysis for your paper, or learn how to do it using a different method. If you want to strengthen your public speaking, join Toastmasters® or cultivate other opportunities outside of your institution by reaching out to your various alma maters and offering to speak to current students. You can see where pursuing these new skill sets can be multifunctional across the skills spectrum. For example, if you are in fact teaching, you are also communicating and presenting, and possibly mentoring.

If you are lacking leadership experience, you can join the postdoc or graduate student association. These groups can be found at the department, school, institution, regional, or national levels, including professional or scientific societies, affinity groups, and the National Postdoctoral Association. As a current and former member of several of these groups, I know they are constantly recruiting for leadership and support roles. If you are interested in science policy, there is plenty of opportunity there, especially in today's complicated scientifically averse political climate. Trained PhDs getting into policy or engaging in the local, national, or global dialogue is vitally important.

As a result of your extensive PhD and postdoc training, you definitely have writing and (maybe limited) editing experience, but you can always get better by diversifying your skill sets. Also, I can attest that while I no longer do research, I am still writing all the time. You can bolster your writing credentials by volunteering to be a guest blogger, becoming a social media content creator, or designing your own website that shares accurate and trustworthy information. You can gain editing experience while raising your professional profile as a reviewer for journals and professional conferences. It may be as easy as offering to help your mentor with a review they were invited to do. There are also freelance writing or consulting opportunities available to you as well. Regardless of what you do to enhance your skills and strengthen your credentials, be sure to do it well, with integrity, professionalism, and authenticity.

6
Path to Independence

Understanding Your Priorities

On your path to independence, all anyone really wants to know about you is whether your research is first rate and whether you can also be a brilliant scientist on a team with other brilliant human beings doing science. No matter what you ultimately do in your career, you have to gain and master many skills not directly related to your scientific training. However, if you are focused and efficient, there are ways to build your professional credentials with relevant experiences. Once I decided to one day run a postdoc office, I realized I needed more experience in four areas: leadership, teaching and service, program management, and writing. I was determined to systematically gain these experiences and take time to reflect on past achievements. Based on my personal experiences, I developed a practical approach to identifying and leveraging experiential opportunities to bolster credentials.

Identifying the most important career priority is a difficult proposition. But in my mind, I think you will be ahead of the game if you spend your energies becoming a respected, recognizable, and influential expert in your field. You can accomplish this by cultivating and maintaining a vibrant network of current and former schoolmates, lab mates, and future employers. To improve your chances of career success in academia or elsewhere, you should be prioritizing research progress where you are publishing and presenting your work. You should also stay visible by keeping in touch with others and building your network. You should go to conferences and meetings, large and small. You should also be open to new ideas as well as unconventional or non-research-related opportunities. Ultimately, your career priorities as a postdoc involve framing your training, building your reputation, creating your vision, and preparing for the next step.

Framing Your Training

It is so interesting how scientists spend years learning a systematic process of observation, research, hypothesizing, experimentation, analysis, and sharing. However, when facing a problem unrelated to research, scientists

FIGURE 6.1
Framing the different phases of your training helps to gain perspective on the transferability of your experience. Image provided by the author.

abandon the scientific method and undertake a haphazard approach that they are unfamiliar with. Scientists need to apply the same rigorous scientific process to their career advancement and professional development as they do to their own research progress. This is not how you would approach your science, so why would you use an unfounded approach in other parts of your life?

It is very important to understand and put your training in the proper perspective (Figure 6.1). Getting a PhD creates opportunity and potential. It hones critical thinking and problem-solving skills. The postdoc, if and when you do it, helps refine your research and professional skills while developing you as an independent investigator. This is really important to remember, that you are on a path toward independence. It is not just in the science or research sense, but you are truly an independent-minded questioner of your surroundings. Ultimately, your training offers a systematic process that prepares you to select the correct problem, critically examine that problem, thoroughly analyze it, and eloquently communicate its solution. If you look at your training holistically, it is basically the mastery of the scientific method. You have spent years and years perfecting this methodical approach to your research. What I want you to do now is realize that you need to use this tool on any problem that you encounter. This is especially true when handling your career and professional development. The attributes and skills that result from extensive PhD and postdoc training in the scientific method are desirable qualities for anyone in any field who is looking to advance to the next phase in their career.

Your Training Efforts

You should have at least four fundamental goals for your PhD and postdoc training. The first is to gain independence through research progress, funding if possible, and collaborations. You can further your independence through leading or supervising others in the form of mentees, peers, and technicians. The second goal of your training is to build a professional identity through networking and fostering relationships alongside having a reputation as an expert in the field or a particular set of techniques. The third goal should be for you to identify a vision for the future that involves both your research and your career. This encompasses your immediate next steps during your postdoc as well as your long-term career plan. Finally, the last of your big-picture postdoc goals is to get a job. But not just any job, it should be one that you find challenging yet is also a satisfying mix of your skills, interests, and values.

After making the necessary efforts to gain additional experience and grow your skills, it is vital to think about how you will frame it in the correct context for current and future use. In doing so, you firmly demonstrate your interests and your willingness to seek out whatever essential training is desired to succeed and advance. This also shows that you have the skills and ability to manage multiple priorities and projects at a time while improving your self-awareness and diminishing your blind spots. In addition to the *actual* experience you received, you have actively developed interpersonal communication, leadership, and strategic thinking skills along the way.

Importantly, I believe the rigorous scientific training you receive during your PhD and postdoc experience helps you discern your career development while building intellectual depth. Everything being equal, effective communication, professionalism, leadership, and continuous learning are the skills that will set you apart from your current peers and your future competition. Early awareness and assessment of your acquired skills give you an advantage not only in navigating your training but also in choosing the right environment to pursue your career. Thus, in the scientific enterprise, you should aim to be trained as a complete person who is capable of effective decision-making, leadership, and independence. I repeat, these are quite desirable characteristics for anyone seeking to advance in any field.

Building Your Reputation

You will generate more opportunities to enhance your reputation and grow your experience by being positive and strategically saying yes. Offering to help out and doing more than just concentrating on your own limited sphere of experimental results establishes you as a resource for others. As you progress, you will find that you have more opportunities coming your way than you can manage, so use your people skills and delegate or share some of them with your colleagues. Once you determine what options are interesting and suitable for you, you will be less likely to miss any unconventional yet relevant opportunities. You can also maximize your efforts by seeking experiential activities within your own research environment. For instance, integrating aspects of your research training that allow for honing your communication, leadership, and collegiality does not take away from your research progress. In fact, you are expected to do these things by participating in regular interactions across the lab, department, or collaborations. You should strive to remain visible in the department by talking with your colleagues, mentors, and faculty members about more than your science.

I will repeat throughout this book that you need to connect with people, speak with others, and continue to broaden your network. However, you should not be compartmentalizing your professional network from your personal network. You should include your family as well as the people you know from school drop-off, place of worship, or the library. For example, I leveraged a random interaction that my mother-in-law had while at her hairdressers to befriend the Director of the Office of Intramural Training and Education at the NIH. Your network should also comprise people you pass in the hallway every day, the lab next door, and surrounding labs and offices too. You should make it a point to know the administrators in your department. This was one thing that was very helpful for me because I quickly realized that they are the gatekeepers to and calendar managers of important faculty and senior officials. Anytime I needed something, be it a quick meeting or a favor, I knew whom to go to. I did not go to the faculty or to the chair of the department directly. No, I went to their administrator and that is how I got things done. And so can you.

Consistently being active and visible within your research community helps establish and reinforce your reputation while allowing others to get to know you and to possibly advocate on your behalf. In this day and age, your community includes the social media realm as there are many active micro-communities found on Twitter, Instagram, and LinkedIn (and more) that facilitate online networking. You have to be serious and genuine about this to make sure you are not being perceived as a show-off or flippant. The simplest way to show that you have some experience is to reframe your hard-earned past and present accomplishments in the context of future endeavors. Further visibility can come from joining a well-recognized group, association, club, or activity. The goal is to contribute immediately, learn teamwork, and potentially rise to leadership. A more difficult but equally rewarding path is creating your own opportunities or starting something new. However, by addressing a universal need or filling a relevant niche, you demonstrate strategic decision-making, initiative, an understanding of stakeholders, and independence.

You can increase the impact of your labors and further expand your network by being generous and sharing credit. Including your peers and colleagues in your growth and development process helps create and foster a community. You may also receive better engagement and guidance if you share the relevance of your efforts with your mentor on a regular basis. Remember to harness the expertise and connectedness of administrative support staff to facilitate access to people and resources. Additionally, you should also not be shy about reaching out to leaders in your field and developing professional relationships with them.

Build Transferable Skills

It is also very important to recognize that you already have A LOT of transferable skills. Sometimes you just have to literally translate what you have done to give context to the work that the next stage of your career is requiring of you. I am positive that if you thought long enough, you could recall that you had planned and organized events, including things for the lab, postdoc association, or greater community. Maybe you had led an event for a conference or at your church or local library. These experiences demonstrate a spectrum of skills and you do get credit for all of them.

I like to point out the fact that no one has done all of their research by themselves in a vacuum. It takes a wide range of transferable skills to do research. There are many times when you need to perform some requisite task, and it is something that you might have little to no experience with. These occasions should serve to remind you that your teamwork and collaboration skills are essential to your research enterprise. If you have ever worked with others and accomplished a goal, you have relationship-building, networking, collaboration, and teamwork skills. Regardless of whether you are a leader or just a contributor, in working with others, you will play a vital role on the team and possibly even influence the process. In fact, you may even get to provide mentorship and supervision for several of the people on the project.

One of the most straightforward ways to demonstrate your project management skills is to leverage what you did to manage the publication process. Many of you reading this book may have published or will publish a paper, especially as a first and/or corresponding author. This process, without you knowing it at the moment, is essentially the definition of project management. You probably oversaw the entire research project, from the design stage to the planning and execution of the overall project all the way down to performing the individual experiments. Throughout the process you managed and communicated deadlines for the research (i.e., authorship) team, even supervising personnel and managing collaborators. You coordinated the full course of writing, editing, revision, and submission. Maybe you even communicated directly with the editor of the journal as you negotiated authorship order while wrangling co-authors to respond to proposed revisions. You led the whole process from ideation to drafting to acceptance to publication and finally to dissemination. That is project management!

You may even have familiarity with aspects of laboratory budgeting, inventory, and workflow. You probably have a very good understanding of time, resource, and personnel management through your supervisory, mentoring, and project management experience. Many of you have been teaching and training high school, undergraduate, or graduate students in the laboratory and classroom. From this, recognize that you have leadership, service, and community-minded skills and experience even though you may not have been given an official title or role. All of these are very valuable and quite transferable though it may take some reflection to frame them appropriately for current and future pursuits.

There are many additional transferable skills inherent in PhD and postdoc training. You may not realize it, but you have very well-developed relational and decision-making skills. In your training you have taken a universal, open-minded, and unbiased approach to problem-solving, all through the use of the scientific method. You have an impressive level of comfort with ambiguity since you are not used to knowing the answers ahead of time. You have had to create new knowledge and maybe even invent new methods and approaches to address your research question. You are proficient in flexible thinking as there is a built-in expectation of failure (before success) in science since you are always trying to disprove the null hypothesis. You have an uncanny ability to pay attention to the smallest details while maintaining a big-picture perspective since you are usually involved from ideation to analysis to communicating the impact of the work. Finally, you have experience in dealing with difficult people, which is unfortunately very common in academia.

Preparing for The Next Step

Because of your specific set of skills and your scientific education, you are always assessing your landscape, training, and self. You will have to continue convincing yourself and others through open-mindedness, storytelling (about yourself and your career), reflective questioning, and contact points. When others have the chance to meet you and hear your story, they can be motivated to be your advocate. Regardless, the three phases of training – career advancement, professional development, and research progress – are not passive. You have to be strategic and active to achieve them. When discussing career opportunities, I have to acknowledge the saturation of career information including non-academic pathways that exist today that were not available during my training. This can be overwhelming. It is key to keep momentum given the intense research demands placed upon you. You do not have to do everything at once, but just be doing something all the time. However, no one thing should be done at the cost of another.

Structuring Your Preparation

Ideally, it is paramount that you are allowing yourself enough time to make preparations for your career transition. Unfortunately, the reality is that I have met postdocs whose appointment or employment was terminated literally the next day. Can you imagine it being your last day as a postdoc, and it being the first time you have reached out to anyone for help? Regardless of your situation, if you come to me with a year or day left on your fellowship, my approach would be similar with a primary aim to help you gain some perspective. Which will still take time. You should reach out for help as soon as you recognize that there is a transition, a change in your status, or a trajectory correction on the horizon. Whether your issue is temporary or permanent, start seeking advice from mentors, peers, and trusted colleagues. Your faculty mentor can be a great resource, assuming you have been cultivating a professional and mutually respectful relationship with them. I do realize that they may also be the precipitating reason for your decision to start your preparations, for good or bad. Chapter 7 goes into much more detail regarding career transition readiness.

When you give yourself enough time and your timeline is on the scale of months or years, you can take the time to survey the landscape of the next phase of your career. In addition to furthering your research, you should be learning the language and expectations of the next sector (academe, industry, consulting, etc.) you are interested in moving into. You can do this by seeking training in such areas as the business of science as well as the administration of research. No matter what your goal is, my advice is to address any gaps you have (or perceive that you have) that may hinder a quick transition. Appropriate training may be difficult to find, but it is not like you need another degree. You just need to do enough to put yourself on the right path where you begin to recognize your transferable skills and how to translate them into a new career context. As you continue to prepare for your next steps, you can take your training as a scientist and leverage your skills not just for the next job but also for the transition itself.

It can be very difficult to balance your efforts. There is a whole industry around time management and multi-tasking. This is where I struggle the most, prioritizing my effort and time. I will not tell you that you need to spend 20% of your day or week on professional development, 15% on career advancement, 55% on your research, and 10% on stimulating pursuits. I have found that the breakdown of your effort (and priorities) changes regularly due to scheduling, availability, and interests, to name a few. The advice I give, and followed as a postdoc, is to integrate as much of what you are doing within your usual research routine. Because of this, you may also have to redefine the scope of your endeavors to include such efforts. Ultimately, none of your professional efforts should be compartmentalized; they should cross over and flow into each other. Specifically, your research existence should intersect with your career advancement, professional development, and stimulating pursuits. The reciprocal is also true as these activities all connect back to and inform your research. You have been trained as a scientist, as a researcher, to solve problems using the scientific method. The idea here is to basically integrate one problem-solving activity into another so that you are efficiently managing yourself, your time, and your effort.

Part 3

The End

7
Career Transition Readiness

Knowing When You Are Ready to Leave

When considering a job or career transition, there are many things you need to do to prepare yourself. You must develop new skills, hone established skills, and leverage as many career development opportunities as you can. You should try to enhance your professional image and visibility through strategic networking and informational interviewing. When networking during a transition, you have to be careful in how you frame your reason for changing jobs as well as describing the skills that will help you when you land. One of the most common things I share when you are in transition is make sure that you are moving *toward* something rather than running *away* from something. This is especially true when you are recontextualizing your experiences for a new audience. It is not that you are leaving academia but that you are taking the next logical step in your career and moving toward something that you are interested in pursuing. This next step sets you on a path where your skills, interests, and core values converge. Another thing I talk about all the time is make sure that you are networking. You should always be trying to continue building up and staying active with your network.

Navigating an Unknown Process

In the next few paragraphs, I will share how many of you are, or will be, inefficient in your job search or transition. (I hope this book helps to maximize your efficiency.) I mentioned in Chapter 6 how you tend to abandon the scientific method at the first sign of non-research-related turmoil. In doing so, you make career decisions without data, evidence, information, or direction. Even worse, you make decisions based on bad advice, erroneous assumptions, fear, frustration, and desperation. You also tend to do the process backward by sending out applications before you have reviewed or polished them. Many of you put off the impending transition as long as possible and end up starting the process almost too late. As shared in Chapter 3, you really should start your postdoc training with the end (i.e., career trajectories) in mind. Regardless, you should try to give yourself anywhere from 6 to 18 months to fully engage in your job search or career transition.

To pile on, you are also inclined to make several unforced mistakes that can hinder your job search. Most importantly, you do not talk with your faculty mentors about any of the processes, unless it is about an academic career, and even then, their involvement can be minimal. You are also not reaching out enough to make connections and ask questions. You have a habit of being too focused on the wrong things such as negotiating offers before you have submitted any applications. You are also worried about publication impact factors rather than identifying and honing your transferable skills. Some of you are so concerned with beating an applicant tracking software (ATS) that you have not met any new colleagues in the field you are hoping to transition into. There are a few other oversights that you make that impact your job search that you might not even know you are doing. You have little to no online presence, whether it is a page on your lab's website, a LinkedIn profile, a Twitter account, or a science-focused social networking site such as ResearchGate. You also do not prioritize career and professional development workshops or opportunities until it is too late.

Some of the biggest frustrations I have observed with regard to your career transitions are all of the mysterious unknowns. You eventually come to realize that this is the first time you have ever done a *real* job search. You may find that there is also a general lack of institutional or mentor support for career exploration, advancement, or transition for postdoctoral trainees. You are not confident in sharing your stories or framing yourselves, your skills, or your experiences. You are unfamiliar with the job search or application process nor are you sure how each component of the process is relevant. You need to learn best practices for career exploration and not myths or singular experiences. There is a lot more for you to address than you first realize.

The challenge that postdocs face when successfully transitioning is proper understanding. You can overcome this by educating yourselves on the process and expectations thereof. You do not want to be unprepared. As you proceed through your research training, try to always be working toward *something* while maintaining flexibility to pivot or revise your trajectory. From a trainee's perspective, you know academia, you mostly know the ins and outs as well as the expectations of that environment. Once you decide to transition to a position beyond your postdoc, there are many things that you do not yet fully grasp. This is not so much a mistake, since it does not really keep you from transitioning, but it is a barrier that you are generally unprepared to face. You are not really sure what you need to be specifically preparing for, but remember your training (and transferable skills) because you are used to ambiguity in your research. Why not embrace the uncertainty in your job search as well? To struggle through that, you should always be working toward a goal while remaining nimble and flexible. You may recall in Chapter 6 where I outlined that your scientific training has prepared you to make transitions. It has prepared you to do what is necessary to make the transition out of your postdoc training and into any independent position you choose. Just remind yourself of what you have been trained to do – discover something new, work through ambiguity, master it, and then use it to your advantage.

Regardless of how you arrive at knowing when you are ready to transition, you should never be reacting out of fear. It is important to contextualize your transition as moving toward, pivoting, or growing into a more suitable situation. As outlined in Chapter 4, many of the initial reasons you might be considering a transition from your postdoc may come from a space of anxiety, negativity, doubt, or even fear. You may have even come to a certain realization that the academic routine or faculty career is no longer for you. The reason could be as simple as a changing or broadening of your skills and interests. Nevertheless, you must recognize that for what it is and try to pivot into something that is a good fit for you.

Activating Your Network

I will continue to repeat this again and again, activate your network, and continue to cultivate that network. What I mean by activating in this sense is that you need to notify your network when you are going on the market or that you are currently on the market. This is one of the quickest ways to find positions and hear about new opportunities, just reach out, and reconnect with people that you know. Tell them you are on the market but do not necessarily ask them to do your search for you, only if they hear or know of anything. While this may be a soft or passive request, it will plant a seed for future reference and your network will likely send listings your way. In addition to activating your network, it is important to cross-train, ensuring what you are doing is multipurpose and who you are meeting are multifaceted, thereby making the most out of your efforts. As you will read later in this chapter, if you are going to do something research related or career related anyway and it is in your calendar, make sure you are taking advantage to do several other things while you are there.

Common Ground

When talking about developing your network, I like to dive into three multifunctional areas: establishing common ground, identifying contact points, and integrating your networking activities. Establishing common ground is very important, and it helps bridge the differences you have with a person or group of people that you are speaking with. A former colleague shared a simple way to do this. They suggested creating a short script that you can use while briefly introducing yourself. The script basically has you saying:

> The _RESEARCH / SCIENTIFIC / PHILOSOPHIC_ interests or challenges I share with you are _X, Y, and Z_. Of these things, I have learned that _THIS_ happens because of _SOMETHING I HAVE DONE_. Through my research or experience, I noticed that _A VERY INTERESTING THINGS OCCURS_ because of _THIS EVEN MORE INTERESTING THING_. More specifically, I would like to know _THIS_.

You can just fill in the blanks (or in this case, the emphasized phrases) mad lib style. Thus, creating a relatively easy and memorable way to introduce yourself while quickly establishing a foundation of mutual understanding. It also allows them to process and connect the information you are sharing with their experience. You might even get a few follow-up questions for clarification or further details. Then, turning your attention to the listener, you ask them, "So, what brings you here [to this gathering]?" Before you know it, you have started a conversation and begun to network at a deeper level. For internationals, you may have to have these conversations as non-native speakers of English. While there are some potential communication difficulties, I am constantly amazed at how gracefully you, as international postdocs, face this language barrier. Instead of worrying about your accent or making a grammatical mistake, you should concentrate on how your listener is absorbing the details of your story. Any spoken errors are either ignored or quickly forgiven by the listener for the sake of establishing a meaningful connection.

Contact Points

When you are talking with people and after you have established some common ground, you will find that as you continue, you immediately begin to identify contact points or potential points of commonality. You are at the center of a vast system of contacts and common points and these networking conversations usually revolve around your (or their) education, PhD training, postdoc experience, or a combination of thereof (Figure 7.1). When first meeting someone, we tend to want to know where the other person is from, either geographically or scholastically, where they went to school, or got their training, followed closely by who they worked with and/or studied under. Academically speaking, you are an alum of your postdoc, graduate, and undergraduate institutions. Some conversations may go as far back as high school. Because of this multi-layered alumni status, you can leverage thousands of potential points of commonality across faculty, students, staff, trainees, and beyond. And we are only just taking into account your academic research experience thus far.

The deeper, broader, and further you go, the more people and places you have to connect with. When connecting with people, you want to remember

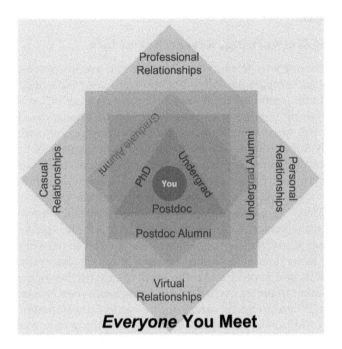

FIGURE 7.1
You are at the center of a vast multifaceted network. Image provided by the author.

that, in addition to your academic experience, you have professional, personal, and virtual relationships. I must acknowledge that the COVID pandemic has changed or even disrupted how we meet people and network. While it has been an overall difficult situation, it has afforded us a common experience from which we can derive shared topics of discussion. In addition, COVID has expanded the ever-growing community of people you have interacted with but have never met in person. It is also important to recognize that your casual as well as your close personal relationships are potential fodder for networking and contact points. If we follow this to its logical end, theoretically, anyone you meet could be quite helpful to add to your network. Remember to always be on your best behavior so you can take advantage of meaningful chance meetings or conversations. I would also note, these layers of contact points are also fantastic prospects to mine when interviewing, whether informational or on the job market.

After meeting someone, establishing common ground, and identifying contact points, do not forget to follow-up with them to further solidify and foster your budding connection. It is not about getting value or information from somebody else; it is actually about providing something worthwhile to others. That value ends up returning to you many-fold if you strategically invest the time and effort early and often.

Integrating Networking, Research, and Life

The most efficient way to network, whether you are introverted or intimidated is not to compartmentalize but integrate your networking. It is not something to be separated from your everyday routine. When networking, or doing anything regarding your career and professional development for that matter, you want to make sure that everything you do serves multiple purposes. By doing this, you begin to unify diverse aspects of your training and share your interests. For example, if you go to department seminars and functions (many of you are required to do so anyway), introduce yourself to other attendees and do not always sit with the same people or even your lab mates. Many departmental seminar series invite famous and up-and-coming scientific leaders to give talks. Be sure to ask the speaker questions and follow-up afterward. This is a great way for you to network within the department as well as outside of it because it is rare that trainees do this. When possible, you should make the most of any further opportunities to meet the speaker in person, whether it is a luncheon, social event, or one-on-one meeting. When attending career or professional development workshops or panels, in addition to potentially learning something new, make sure you are taking full advantage by asking questions and meeting the speaker(s) as well as the organizers.

EXAMPLE OF FOLLOWING UP WITH A SPEAKER VIA EMAIL

Hi Dr. XYZ,
I really loved meeting you and Rayanne today and greatly enjoyed attending your presentation. I look forward to implementing your cover letter and resume tips in my own job search.

I was wondering if your offer to assist in further developing my materials was a genuine one. I am very interested in persuing research-oriented industry jobs and experiences. I greatly admire how you seek to help PhDs transition from academia to industry, a transition I now understand more fully thanks to your talk today.

I look forward to hearing back from you regarding the opportunity to improve my application and thank you again for your time today.
All the best,
Lucy

Even if you are unable to attend a specific seminar or workshop, you can still reach out to the speaker, whether it is days, weeks, or months after the fact. Believe it or not, you can still reach out to a speaker and say, "Dr. So-and-so, you were at my institution and gave a seminar 6 months ago. I am sorry to have missed your workshop on XYZ topic at ABC University but I wanted to follow-up and ask you a few questions." I also recommend going to job fairs, even when you are not actively on the job market, to introduce yourself, exchange contact information, and possibly set a time to meet later. You can incorporate networking into your calendar and schedule specific times to check in with your professional network. You can also bring a trusted colleague to ease the pressure of official networking events. Just do not overdo it – you are allowed to meet people one person at a time. Regardless, these are all fairly low-risk, high-reward settings and you have a chance to make a really good impression as well.

It is important to continue to reach out to and connect with other people. However, you must understand that you may need to do more than just network to expand your contact list. You also need to do informational interviews, which are more targeted professional and in-depth conversations with people who are in positions to which you may aspire. Because of these discussions, you are much more prepared to speak the language while furthering your understanding of the new career path or sector you may be entering. Subsequently, you will gain more contacts and referrals to more people you can meet, potentially leading to increased interest in or elimination of future options.

In addition, you have a peripheral set of contacts that exist outside of your mentors, colleagues, alumni groups, and friends which can be called a "stealth network." Your stealth network includes people at your places of worship, parents at your child's daycare, or even your barbershop or hairstylist. They may be the people you sit next to at the coffee shop, work out with at the gym, or compete against in your basketball, golf, softball, or kickball rec league. While being very careful on public transportation (which I use every day), you may strike up a conversation with someone you see on a regular basis or has a similar commute schedule. My point is when you are open and transparent in your conversations, then you are staying visible while interacting with authenticity and sharing your interests with other people.

Having "The Talk" with Your Mentor

You need to actively manage your local support systems, especially your faculty mentor. However, many of you may not feel fully supported by your faculty mentor or your lab environment. You may be afraid to discuss any career path with your faculty mentor for fear that you might be asked to leave prematurely, will be supported less, or that your contract will not be renewed. If you do not have a supportive mentor who helps to connect you to their network, then you can feel quite trapped. This is particularly true if your career trajectory takes an unexpected turn and you have to make a change to another lab or from academia altogether. Nevertheless, the best source of information and support for your career transition remains your faculty mentor. You should continue to have meaningful, ongoing conversations about your research and your career. What you can do right now is to fully engage with your mentor, not do this alone, and keep those lines of communication open. Chapter 5 covers several avenues for you and your mentor to foster a productive relationship including implementing an IDP, setting expectations, and respecting boundaries.

As you progress on your path to independence, it is vital that you allow them to come along with you on your journey. They can help you distinguish whether or not you are actually ready for the next stage of your career by sharing their knowledge of what it takes to successfully transition. If you bring them into the discussion, your mentor can assist in discerning the type of role you want in the field or sector into which you are considering a transition. They can provide valuable insight into where you may be best suited to work, and what it is like being an assistant professor, or they can introduce you to a colleague of theirs who is a senior scientist in biotech. They can also help you define a realistic job search timeline that leads to a seamless exit from the lab. Your mentor's advice and insight will be critical when deciding to start your search. If you go too early and your projects or fellowship are not yet complete, you risk not being able to commit to an offered position. Similarly, if you launch a search too late, you might be without an offer in time for your expected exit. Even if you timed it right, you may not receive interviews or offers and will need to go on the market more than once. Each of these scenarios takes time, effort, and a network of support.

You may also be interested to know that once you are committed, your job search, regardless of sector, will become a second full-time job that you will have to manage. It will be infinitely more achievable with the support of your mentor and lab mates. The job search timeline varies depending on your endpoint, but generally from application submission to job offer, the academic job cycle can take up to a year while a non-academic search could last as long as six months. In that time, not only will you be continuing experiments and submitting manuscripts, but you will also be expected to manage multiple applications, gather your references, and navigate multiple rounds of phone, virtual, and onsite interviews. The success of this process relies, in part, on negotiating time away from the lab and your projects with at least the knowledge, hopefully full backing, and ideally actual assistance, from your faculty mentor. Stay open-minded to what your mentor can do for you as they can be your greatest ally if you give them the chance. The worst thing you could do during your postdoc is to become isolated. You do not want to be alone and on the verge of burn out when making these decisions.

Launching Your Job Search

You need to be clear about when you are launching your job search. Once you submit an application you are signaling to the universe that not only are you initiating your job search but you are prepared to leave your current role. When that hiring manager reads your application, they expect you to understand their mission, speak their language, and know their approach to science. As part of your postdoc tenure, you will have to decide what path to pursue. What makes this extraordinarily tough is that whatever path you choose likely requires a similar skillset, at least initially. It takes a team of mentors and trusted peers to talk this through. To me, it comes down to the pace of life, work/life considerations, type of research, and whether or not you enjoy grant writing and competing for funding. Choosing a career path and when to pursue it are very personal decisions where you have to line up the pros and cons and understand where the intersection of your skills, interests, and values lies. All while acknowledging that there are countless different options across the scientific enterprise and beyond.

The ideal job search process is a multistep sequence. If you have the foresight, you give yourself a timeline of at least a year when you know you will be on the job market. The longer the better, but at least a year. The job search is much more self-reflective and interpersonal than anybody gives it credit for. The first thing to do is some fairly in-depth self-reflection as laid out in Chapter 3. Begin to think about what you like to do, what you are good at, and what your values are. Then follow-up with your network contacts who share similar attributes. Identify what resources you have institutionally, in the lab, at your *alma mater*. You must recognize that you have many resources just within your grasp, all you need to do is find and use them. When you have people helping you, ask them what the current job trends are, what they have seen recently, and where postdocs with similar training have found success. After gathering this important information, you can begin to refine your vision for the job best suited for you.

Potential Job Search Sequence

1) Establish a timeline
2) Activate your network
3) Self-assessment and reflection
4) Identify skills-interests-values
5) Leverage institutional resources
6) Informational interview
7) Hone application materials
8) Submit applications
9) Interview prep and performance
10) Negotiate and accept job

As you visualize your ideal job, you should share that vision with people who have jobs that are very close to what you are describing. Through informational interviewing, get their insights and begin crafting your application materials. At the same time, you should be passively collecting interesting job ads and opportunities. Then, you go to a career counselor, or a coach, and get feedback on your materials to make them stronger, more relevant, and representative of a viable candidate in your desired career path. Give yourself time to digest what is out there and analyze job advertisements to learn how to read between the lines. Chapter 8 summarizes how to go about putting together a practical application through a series of revisions, peer review, expert critique, and then beginning to apply. You must be very strategic, focused, and targeted. Please be realistic, however, because your first job out of postdoc may not be the absolute ideal one yet but will hopefully put you on the right path. This is but the beginning of the rest of your career.

If you are leaving your postdoc under bad circumstances, I remind you of my earlier advice, to make sure that you are coming from a mindset and a perspective of future-focus and optimism. You have to take care that you are strategically planning out your next moves. You should be pivoting into the next logical step rather than looking back from a place of fear or frustration. It is okay to feel that way, but you cannot have your decisions dictated by distress. You must tap your resources to help you move past your struggles by reaching out for help, activating your network, and cross-training, thus creating positive and forward momentum.

8
Some Advice on Application Materials

Applications and Job Postings

The Struggles and Strategies of Application Material Prep

In making the transition from postdoc to job applicant or candidate, most of you tend to be too narrow in the way you think about and draw attention to your skills and experience. Additionally, you may be inclined to prematurely select yourselves out of opportunities for baseless reasons. Let employers tell you no, besides you can always choose to remove yourself from consideration if you receive interviews or offers. When drafting your application materials, you are also apt to write them for yourself and not for the hiring committee to understand while also submitting your application with no input or feedback. It is nearly impossible to craft a compelling application that no one else has read or critiqued.

As postdocs, you may find such career transition topics and materials to be overwhelming and confusing. Thus, you are often mystified and even paralyzed as to how to proceed. Using peer review and critique to hone application documents, as well as focusing on self-awareness and authenticity, any training you undertake to better your applications should cover common struggles and effective strategies for preparing you for successful job searches and career transitions. Through my own work in this area, I have found that you often identify struggles, as evidenced above, that are commonly shared across multiple career transition topics, as well as some potentially effective strategies to deal with such challenges.

For instance, when it comes to drafting materials and preparing for career transitions, you will be surprised to learn that most of you have the same concerns, regardless of background, experience, or topic. In general, you are likely to not understand the true purpose of some or all of the components within the application package. You also have questions about each application document's length, structure, order of sections, formatting, and balance of interpersonal and technical content. The best strategies to address such commonly held struggles can actually be universally applied across all materials. When you are crafting your applications, you should have others review your documents, find and model recent successful examples, and mirror the language of the job posting, if possible. Ask to see the materials your colleagues used in their job search. Having examples is helpful, but it is critical to understand what made them effective. Additional advice is for you to always be yourselves, honest, authentic, and direct. You believe that you can do the job, now tell *them* all about it, with concrete examples.

Stretch Positions and Multiple Applications

It is quite alright to apply to stretch positions that may seem *just* beyond your reach or skillsets. Besides, if you wait until you are 100% qualified for a job then two things are likely true: (1) you are now overqualified, and (2) you missed out on many, many opportunities. In addition to being interested, you just have to make sure that you are at least meeting the minimal qualifications and that you have a few of the required technical or soft skills for the job. Networking will definitely help you understand what those stretch jobs entail. You should also highlight the desirable skills that you already have.

Another common issue you may encounter is that a company or institution that you find interesting has multiple open positions listed that you feel qualified to fill. You may worry about coming off as desperate, but if you do not know anyone inside the company or institution, how else are you supposed to get on anyone's radar outside of applying? What you want to do is make sure that you tailor your application materials to those specific jobs but not so over-tailored that it looks like you are a totally different candidate for each one. You can also disclose in your respective cover letters that you have applied to a few other opportunities within the company or institution and hope that you can be considered as appropriate. The employer may also be able to cross-reference applications, applicants, and different positions. To be sure, if they have ten openings, I would not suggest applying to all of them, perhaps limiting your interest to a select few.

Job Postings

Almost everything you need to know about a job is in the posting. Understanding the purpose, style, and general format of your application are the first steps in creating an effective package. While drafting your documents, do not miss a chance to incorporate the information laid out in the job posting. You should repeat the keywords provided, mirror their language, and recognize all aspects of the posting. Most job postings follow a general structure. To start with, it usually gives some level of background about the position as well as information about the employer and its mission, vision, and perks of working there. Of course, it shares minimum research and technical qualifications, education, and years of experience. The posting also contains preferred technical, interpersonal, and professional skills and experience in a specific environment or field. Typically, there are also application submission instructions or a link to an online portal. Please be sure to read and follow all guidelines. If you have any questions, do not hesitate to ask for clarification by reaching out to the contact person, if provided.

Cover Letters

Crafting the Cover Letter

A cover letter is a great opportunity for you to capture the attention of the hiring manager or search committee by providing the proper representative context to your experience while controlling the narrative of your candidacy. While many positions allow you to apply directly from your LinkedIn profile without a cover letter, it is a good idea to have one at the ready and an excellent exercise to help you promote yourself. By understanding the job posting and aligning the cover letter to its content, you can quickly and concisely convey interest and job fit, all in just one page. In crafting and tailoring a cover letter, you must address specific qualifications employers are looking for when they advertise open positions. The cover letter essentially summarizes the CV/résumé with concrete details. But not everything can be included; therefore, you must use representative experiences. This document should be able to stand alone while still complementing other materials. You should draft a clean, clear, concise, and flexible cover letter written in active voice with traditional style conventions that are easy to follow. Considering your reader and your purpose, this document has to be specific to the position with obvious connections.

Generally, the first paragraph of your cover letter states why you are writing (applying to a specific position), the position title and requisition number, and refers to how you learned about the employer or the job whether it was through a previous conversation, meeting, correspondence, or just the job posting. You should convey a sense of enthusiasm about the position, department, company, or sector; use the opportunity to demonstrate you know something about them. They want to know what excites you enough about the position to apply. Keep it short, upbeat, and engaging while avoiding clichés and platitudes.

The following paragraph is where you describe the "big picture". For example: "As a postdoctoral research fellow at Respected Medical School, I have been investigating the effects of X on Y under the supervision of Respected Professor." Start with your postdoc experience and, if relevant, work your way backward to your graduate training describing your research, underscoring its significance, and adding publications. Include other significant experiences, education, and research as appropriate while highlighting the breadth of your expertise. Elaborate on your distinctive qualifications, strengths, achievements, skills, and experiences. The amount of information and what is relevant to include will be determined by the job posting. You need to understand what they are looking for and incorporate details from your training that will directly address their hiring needs. If it is not relevant to the position, you do not need to include it in your cover letter.

You further address your "fit" in the subsequent paragraphs using concrete examples of your soft skills and related technical expertise. You could also weave this information into the experiences outlined above to tell a more fleshed-out story. Demonstrate that your skills, interests, and values are suited to the position (e.g., pace of discovery, use of innovative technologies, collaborative research setting, *etc.*). Show that you know what a job or career in that field entails, stating why you are interested in working for this employer and your interest in this specific type of work. The remainder of this paragraph is used to make a direct connection between you and the requirements of the job, using several key requirements for the job to demonstrate how your experiences and training fit with what they are seeking. This could include specific leadership experience, research expertise, collaborative approaches, techniques, or technologies. The better the match, the better your chances of getting invited for an interview. It should be obvious to the reader that this is the next logical step in your career.

Wrap up the final paragraph by continuing to convey a sense of enthusiasm about the position. Offer that you look forward to discussing the position and tell them again how to contact you. End with a sincere concluding remark and thank them for their time in reviewing your materials.

As you can imagine, the Academic Cover Letter, while very similar in overall structure and content, is expected to be framed in a manner that demonstrates how well-prepared you are to independently run a lab in an academic setting. The cover letter for your academic job package should be about one page and contain roughly six paragraphs. The opening paragraph should include your name, what position you currently have, the location and PI of your lab, and what position you are applying for. It also contains a brief mention of specific motivations. You need to write one to two past research paragraphs that are short and punchy, conceptual, and mostly focus on postdoc research, with a little graduate work, if relevant. Because you are including research findings, you would need to add citations using a short reference format. In your future-plans paragraph, focus on the novelty and impact of pursuing the proposed research. While there is no need to specifically refer to your Research Statement here, you will essentially be giving a high-level summary of it. If you will be expected to teach, or are applying to a more teaching-focused faculty position, you will need to add a short paragraph outlining your teaching approach and experience. Again, there is no need to refer to your Teaching Statement but only share an overview. In the exit paragraph, in addition to the ideas listed above for a general cover letter, include the names and contact information of your references as well as the documents enclosed in your application.

Cover Letter Struggles and Strategies

In crafting your cover letter, I know that you will worry about and be unsure of many things. Some common struggles that postdocs like you have shared with me range from choosing the right experience to emphasize or highlight; to effectively showing interest in a job without being desperate; to providing an appropriate level of detail short of repeating your CV/résumé. You also struggle with the notion that you somehow have to prove yourself or sell your skills instead of keeping to the facts and telling the story of your experience. In addition, I have observed a great many postdocs making unforced errors in their cover letters. The most common being a lack of concrete research context, impact, or examples; instead, you are inclined to use generalized platitudinous phrases like having "extensive experience in," "profound knowledge in," or "a passion for" a specific discipline. It is likewise bothersome when you reference your CV, mention a publication, or name-drop rather than provide the interesting details of your actual accomplishments in the document in front of the reader. I have also witnessed egregious mistakes such as naming the wrong institution, misspelling or making up words, and unnecessarily calling attention to a red flag.

Of course, there are myriad strategies to address much of the above, most being no-nonsense approaches that you can use for all of your application materials (as mentioned at the beginning of the chapter). Though, for cover letters specifically, you can read them out loud to yourself to catch any grammatical or logical errors while asking others to read and review the document. As outlined, it is critical that you read, understand, and deconstruct the job ad making sure to include relevant, representative, and unambiguous experiences. Finally, you should tell an authentic story that addresses fit as well as conveys a focused career trajectory.

Crafting the Résumé and CV

Understanding Your CV and Résumé

A CV or résumé provides the first impression of a candidate to a potential employer. The content of the document should quickly grab the reader's interest, and the format must deliver information clearly and concisely. To make it even more challenging, you need to summarize the highlights of your career in just a few pages. It may also be necessary to have a few different versions depending upon your pursuits. To be sure, a CV is not résumé and vice versa, even though some may use the terms interchangeably. By definition, a CV is curriculum vitae literally meaning the course of one's life work. A CV is a bit more comprehensive as it can include most, if not all, of your academic credentials. On the other hand, a résumé is relatively short, skills-based, and very tailored to the job to which you are applying. Regardless of which one you use for your application, you must realize that this document will only get you an interview or a meeting, not a job offer.

CV, Résumé, or Something in Between

A CV, to me, can be described in three parts: a *Pedigree* that includes your degrees, where you studied, and who you studied with; some *Padding* that fleshes out your academic service; and *Publications* that list your contributions to the scientific literature. In the CV, you provide a chronological account of your academic credentials while giving context to your research experience, and this is used primarily for academic positions. Some may consider this to be the end-all-be-all of your academic existence, so, depending on your relatively early career stage (PhD student or postdoc), a concise version of your CV might be three to five pages. If it was any longer than that, readers may begin to think you are adding extraneous or irrelevant information. While I highly recommend having a "master document" containing all of your experiences and achievements, you must be able to create a concise and focused document to use for applications. The CV basically says, "Look how brilliant I am. How could you not interview someone as smart as me?"

The résumé, instead, includes your *pedigree*, as above, but also frames your accomplishments and service as *proficiencies* and *productivity*. Outside of academia, the tendency to think of soft skills as "padding" and thus less important is unfounded as those same experiences are considered very important and become valued proficiencies. Additionally, as opposed to academia where the sum total of your training is distilled down to your publications, in other sectors, your scientific productivity is measured more broadly. Similar to the CV, a résumé is a chronological account of your academic credentials but is very skills-oriented by including research, technical, and soft skills. Unlike the CV, a résumé is a snapshot of your experiences that fit and are relevant to the job for which you are applying. It is a concise one- to two-page document that gives context to your research accomplishments, ultimately telling the hiring manager that you can do the job. Overall, your résumé should say, "Look how my skills and experience match your job. How could you not interview someone as well-suited as me?"

However, as constantly mentioned throughout this book, postdocs are special in many, many ways. Due to your advanced education, extensive research training, and fluid early career, I have come to advocate for the use of a transitional CV/résumé hybrid document that combines the academic credentials of the CV with the skills-focused experiences of the résumé (Figure 8.1). This hybrid document should be two to three pages and can be tailored for a variety of research and research-adjacent positions in industry, academia, and private sectors.

Purpose Impacts Style

You want to understand what your CV and résumé is and what it can and cannot do because the purpose of this document impacts the style in which you will format it. For example, if you are looking for an academic faculty

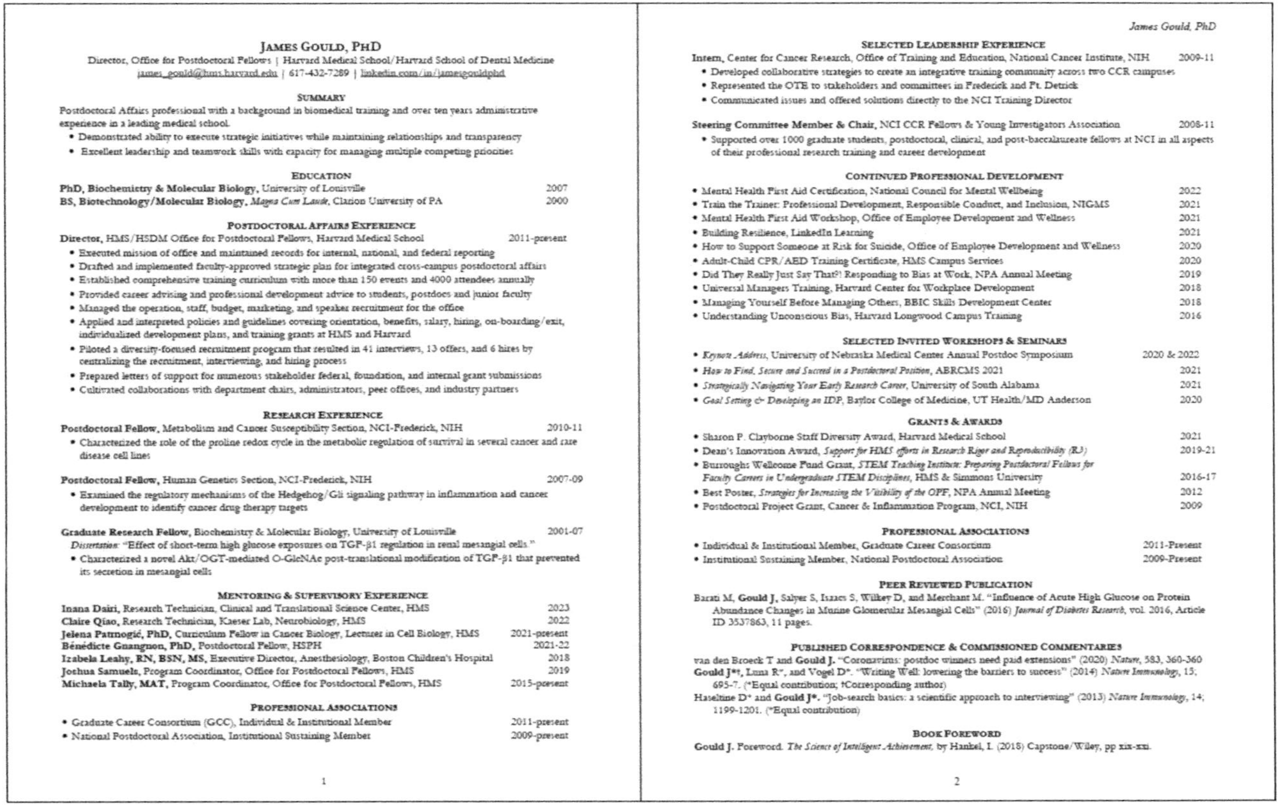

JAMES GOULD, PHD
Director, Office for Postdoctoral Fellows | Harvard Medical School/Harvard School of Dental Medicine
james_gould@hms.harvard.edu | 617-432-7289 | linkedin.com/in/jamesgouldphd

SUMMARY
Postdoctoral Affairs professional with a background in biomedical training and over ten years administrative experience in a leading medical school.
- Demonstrated ability to execute strategic initiatives while maintaining relationships and transparency
- Excellent leadership and teamwork skills with capacity for managing multiple competing priorities

EDUCATION
PhD, Biochemistry & Molecular Biology, University of Louisville 2007
BS, Biotechnology/Molecular Biology, *Magna Cum Laude*, Clarion University of PA 2000

POSTDOCTORAL AFFAIRS EXPERIENCE
Director, HMS/HSDM Office for Postdoctoral Fellows, Harvard Medical School 2011-present
- Executed mission of office and maintained records for internal, national, and federal reporting
- Drafted and implemented faculty-approved strategic plan for integrated cross-campus postdoctoral affairs
- Established comprehensive training curriculum with more than 150 events and 4000 attendees annually
- Provided career advising and professional development advice to students, postdocs and junior faculty
- Managed the operation, staff, budget, marketing, and speaker recruitment for the office
- Applied and interpreted policies and guidelines covering orientation, benefits, salary, hiring, on-boarding/exit, individualized development plans, and training grants at HMS and Harvard
- Piloted a diversity-focused recruitment program that resulted in 41 interviews, 13 offers, and 6 hires by centralizing the recruitment, interviewing, and hiring process
- Prepared letters of support for numerous stakeholder federal, foundation, and internal grant submissions
- Cultivated collaborations with department chairs, administrators, peer offices, and industry partners

RESEARCH EXPERIENCE
Postdoctoral Fellow, Metabolism and Cancer Susceptibility Section, NCI-Frederick, NIH 2010-11
- Characterized the role of the proline redox cycle in the metabolic regulation of survival in several cancer and rare disease cell lines

Postdoctoral Fellow, Human Genetics Section, NCI-Frederick, NIH 2007-09
- Examined the regulatory mechanisms of the Hedgehog/Gli signaling pathway in inflammation and cancer development to identify cancer drug therapy targets

Graduate Research Fellow, Biochemistry & Molecular Biology, University of Louisville 2001-07
Dissertation: "Effect of short-term high glucose exposures on TGF-β1 regulation in renal mesangial cells."
- Characterized a novel Akt/OGT-mediated O-GlcNAc post-translational modification of TGF-β1 that prevented its secretion in mesangial cells

MENTORING & SUPERVISORY EXPERIENCE
Inana Dairi, Research Technician, Clinical and Translational Science Center, HMS 2023
Claire Qiao, Research Technician, Kaeser Lab, Neurobiology, HMS 2022
Jelena Patrnogić, PhD, Curriculum Fellow in Cancer Biology, Lecturer in Cell Biology, HMS 2021-present
Bénédicte Gnangnon, PhD, Postdoctoral Fellow, HSPH 2021-22
Izabela Leahy, RN, BSN, MS, Executive Director, Anesthesiology, Boston Children's Hospital 2018
Joshua Samuels, Program Coordinator, Office for Postdoctoral Fellows, HMS 2019
Michaela Tally, MAT, Program Coordinator, Office for Postdoctoral Fellows, HMS 2015-present

PROFESSIONAL ASSOCIATIONS
- Graduate Career Consortium (GCC), Individual & Institutional Member 2011-present
- National Postdoctoral Association, Institutional Sustaining Member 2009-present

1

James Gould, PhD

SELECTED LEADERSHIP EXPERIENCE
Intern, Center for Cancer Research, Office of Training and Education, National Cancer Institute, NIH 2009-11
- Developed collaborative strategies to create an integrative training community across two CCR campuses
- Represented the OTE to stakeholders and committees in Frederick and Ft. Detrick
- Communicated issues and offered solutions directly to the NCI Training Director

Steering Committee Member & Chair, NCI CCR Fellows & Young Investigators Association 2008-11
- Supported over 1000 graduate students, postdoctoral, clinical, and post-baccalaureate fellows at NCI in all aspects of their professional research training and career development

CONTINUED PROFESSIONAL DEVELOPMENT
- Mental Health First Aid Certification, National Council for Mental Wellbeing 2022
- Train the Trainer: Professional Development, Responsible Conduct, and Inclusion, NIGMS 2021
- Mental Health First Aid Workshop, Office of Employee Development and Wellness 2021
- Building Resilience, LinkedIn Learning 2021
- How to Support Someone at Risk for Suicide, Office of Employee Development and Wellness 2020
- Adult-Child CPR/AED Training Certificate, HMS Campus Services 2020
- Did They Really Just Say That?! Responding to Bias at Work, NPA Annual Meeting 2019
- Universal Managers Training, Harvard Center for Workplace Development 2018
- Managing Yourself Before Managing Others, BBIC Skills Development Center 2018
- Understanding Unconscious Bias, Harvard Longwood Campus Training 2016

SELECTED INVITED WORKSHOPS & SEMINARS
- *Keynote Address,* University of Nebraska Medical Center Annual Postdoc Symposium 2020 & 2022
- *How to Find, Secure and Succeed in a Postdoctoral Position,* ABRCMS 2021 2021
- *Strategically Navigating Your Early Research Career,* University of South Alabama 2021
- *Goal Setting & Developing an IDP,* Baylor College of Medicine, UT Health/MD Anderson 2020

GRANTS & AWARDS
- Sharon P. Clayborne Staff Diversity Award, Harvard Medical School 2021
- Dean's Innovation Award, *Support for HMS efforts in Research Rigor and Reproducibility (R3)* 2019-21
- Burroughs Wellcome Fund Grant, *STEM Teaching Institute: Preparing Postdoctoral Fellows for Faculty Careers in Undergraduate STEM Disciplines,* HMS & Simmons University 2016-17
- Best Poster, *Strategies for Increasing the Visibility of the OPF,* NPA Annual Meeting 2012
- Postdoctoral Project Grant, Cancer & Inflammation Program, NCI, NIH 2009

PROFESSIONAL ASSOCIATIONS
- Individual & Institutional Member, Graduate Career Consortium 2011-Present
- Institutional Sustaining Member, National Postdoctoral Association 2009-Present

PEER REVIEWED PUBLICATION
Barati M, **Gould J,** Salyer S, Isaacs S, Wilkey D, and Merchant M. "Influence of Acute High Glucose on Protein Abundance Changes in Murine Glomerular Mesangial Cells" (2016) *Journal of Diabetes Research,* vol. 2016, Article ID 3537863, 11 pages.

PUBLISHED CORRESPONDENCE & COMMISSIONED COMMENTARIES
van den Broeck T and **Gould J.** "Coronavirus: postdoc winners need paid extensions" (2020) *Nature,* 583, 360-360
Gould J*†, Luna R*, and Vogel D*. "Writing Well: lowering the barriers to success" (2014) *Nature Immunology,* 15; 695-7. (*Equal contribution; †Corresponding author)
Haseltine D* and **Gould J*.** "Job-search basics: a scientific approach to interviewing" (2013) *Nature Immunology,* 14; 1199-1201. (*Equal contribution)

BOOK FOREWORD
Gould J. Foreword. *The Science of Intelligent Achievement,* by Hankel, I. (2018) Capstone/Wiley, pp xix-xxi.

2

FIGURE 8.1
A sample "hybrid" document using the author's credentials that represents a transition between a CV and résumé targeting a position in Postdoctoral Affairs. Image provided by the author.

position, the document you are creating will be a full CV. If you are looking for a second postdoc or a residency, you will be using a CV that includes a technical skills section. If you are going up for an internal promotion at an academic institution, you will probably have to format your CV into a standardized institutional document which is fairly comprehensive and usually inflexible. I must note that, because it is designed for internal promotion, the institutional CV format is not especially helpful when you are on the academic job market and applying elsewhere. If industry research is your goal, I recommend that you draft a document that is more of a CV/résumé hybrid. A transition into a non-profit, entrepreneurial, science-adjacent, or non-science career is going to use a résumé or hybrid of one, two, or three pages at most. If you are submitting a grant or a grant update, you should NOT be using a CV, that will require you to draft a Biosketch (for US federal funding you can use the NCBI's SciENcv system). Please note that these two documents, the CV and Biosketch, should not be treated as equivalent because they convey different information for distinctly different purposes. In conclusion, one size does not fit all and the style with which you write this document is going to change depending on the purpose it is supposed to fulfill.

How CV/Résumés Are Read

After submission and upon receipt, CVs and résumés are processed very quickly. While they may be read multiple times by many people, the very first time they are seen, you have maybe 30–60 seconds to make an impression. What the hiring manager, search committee, or HR representatives are looking for is your academic, education, and research experience (Pedigree). Then they are going to flip over to the back and look at your publication list (Productivity). This is changing, however, due to the extensive use of computers and digital devices to share and view documents, where a reader now just continuously scrolls through the content. If they continue to look at the document, then they open the subsequent middle pages and look at your soft skills, presentations, and anything else you have in there (Proficiencies). Regardless, the three most important things, at the first pass, are your education, experience, and publications (i.e., Pedigree, Proficiencies, Productivity).

This document must be organized logically so that information is accessible and obvious. It also unfolds in reverse chronological order, meaning that the most recent experience is listed first. There should be no extra work required by the reader to find or interpret the details elsewhere. This means that the reader will not have to do any extra mental gymnastics to figure out what you actually mean. The document also has to have uniform formatting along with straightforward and specific headings. To ensure your document can be understood by those trying to hire you, the best thing you can do is have trusted colleagues read and critique it. This goes for all of your application materials.

Organizational and Formatting Conventions

The organizational conventions for the CV or résumé should enhance the readability and not distract from the content (Figure 8.2). Please note that expected conventions can vary across fields, sectors, and regions (foreign

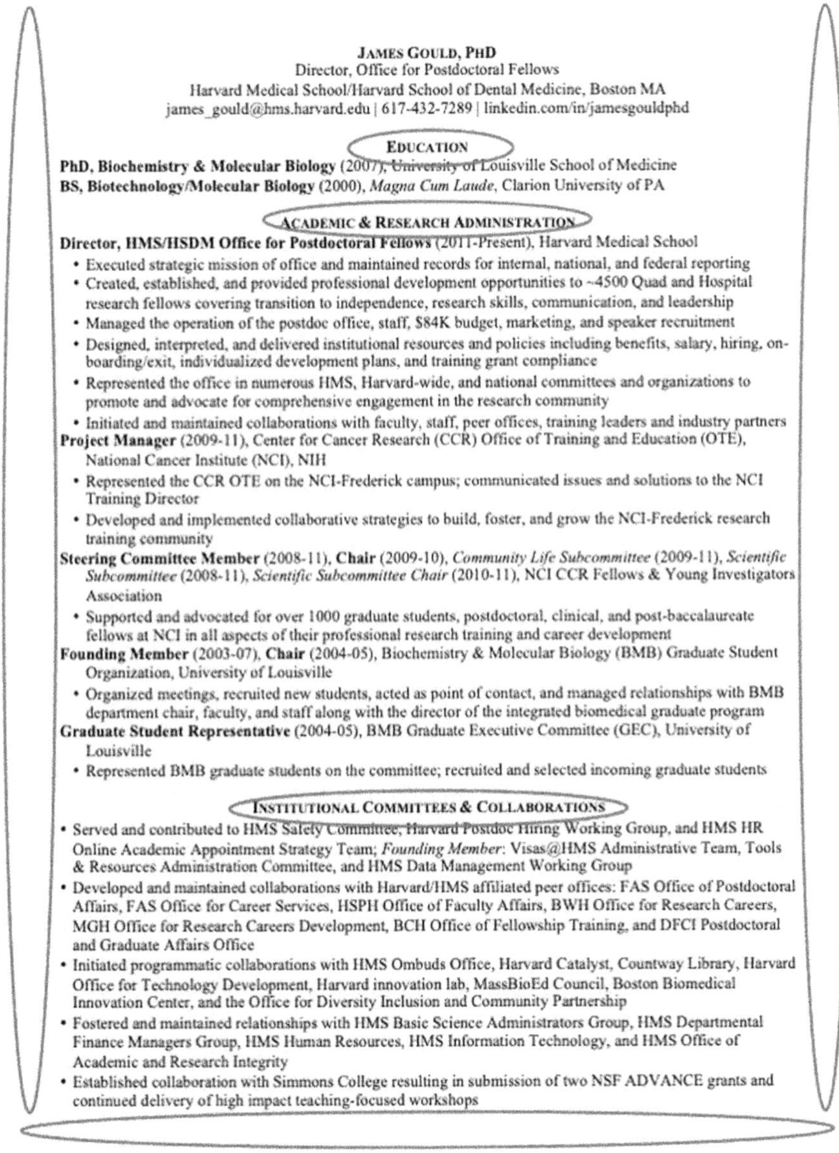

JAMES GOULD, PhD
Director, Office for Postdoctoral Fellows
Harvard Medical School/Harvard School of Dental Medicine, Boston MA
james_gould@hms.harvard.edu | 617-432-7289 | linkedin.com/in/jamesgouldphd

EDUCATION
PhD, Biochemistry & Molecular Biology (2007), University of Louisville School of Medicine
BS, Biotechnology/Molecular Biology (2000), *Magna Cum Laude*, Clarion University of PA

ACADEMIC & RESEARCH ADMINISTRATION
Director, HMS/HSDM Office for Postdoctoral Fellows (2011-Present), Harvard Medical School
- Executed strategic mission of office and maintained records for internal, national, and federal reporting
- Created, established, and provided professional development opportunities to ~4500 Quad and Hospital research fellows covering transition to independence, research skills, communication, and leadership
- Managed the operation of the postdoc office, staff, $84K budget, marketing, and speaker recruitment
- Designed, interpreted, and delivered institutional resources and policies including benefits, salary, hiring, on-boarding/exit, individualized development plans, and training grant compliance
- Represented the office in numerous HMS, Harvard-wide, and national committees and organizations to promote and advocate for comprehensive engagement in the research community
- Initiated and maintained collaborations with faculty, staff, peer offices, training leaders and industry partners

Project Manager (2009-11), Center for Cancer Research (CCR) Office of Training and Education (OTE), National Cancer Institute (NCI), NIH
- Represented the CCR OTE on the NCI-Frederick campus; communicated issues and solutions to the NCI Training Director
- Developed and implemented collaborative strategies to build, foster, and grow the NCI-Frederick research training community

Steering Committee Member (2008-11), **Chair** (2009-10), *Community Life Subcommittee* (2009-11), *Scientific Subcommittee* (2008-11), *Scientific Subcommittee Chair* (2010-11), NCI CCR Fellows & Young Investigators Association
- Supported and advocated for over 1000 graduate students, postdoctoral, clinical, and post-baccalaureate fellows at NCI in all aspects of their professional research training and career development

Founding Member (2003-07), **Chair** (2004-05), Biochemistry & Molecular Biology (BMB) Graduate Student Organization, University of Louisville
- Organized meetings, recruited new students, acted as point of contact, and managed relationships with BMB department chair, faculty, and staff along with the director of the integrated biomedical graduate program

Graduate Student Representative (2004-05), BMB Graduate Executive Committee (GEC), University of Louisville
- Represented BMB graduate students on the committee; recruited and selected incoming graduate students

INSTITUTIONAL COMMITTEES & COLLABORATIONS
- Served and contributed to HMS Safety Committee, Harvard Postdoc Hiring Working Group, and HMS HR Online Academic Appointment Strategy Team; *Founding Member*: Visas@HMS Administrative Team, Tools & Resources Administration Committee, and HMS Data Management Working Group
- Developed and maintained collaborations with Harvard/HMS affiliated peer offices: FAS Office of Postdoctoral Affairs, FAS Office for Career Services, HSPH Office of Faculty Affairs, BWH Office for Research Careers, MGH Office for Research Careers Development, BCH Office of Fellowship Training, and DFCI Postdoctoral and Graduate Affairs Office
- Initiated programmatic collaborations with HMS Ombuds Office, Harvard Catalyst, Countway Library, Harvard Office for Technology Development, Harvard innovation lab, MassBioEd Council, Boston Biomedical Innovation Center, and the Office for Diversity Inclusion and Community Partnership
- Fostered and maintained relationships with HMS Basic Science Administrators Group, HMS Departmental Finance Managers Group, HMS Human Resources, HMS Information Technology, and HMS Office of Academic and Research Integrity
- Established collaboration with Simmons College resulting in submission of two NSF ADVANCE grants and continued delivery of high impact teaching-focused workshops

FIGURE 8.2
An example page highlighting typical organizational and formatting conventions. Image provided by the author.

and domestic). Generally, the heading and contact info traditionally include your name and associated degrees (i.e., James Gould, PhD) in big bold letters at the top followed by an address, phone numbers, and email. For those of you that may have changed your name, perhaps converting your maiden name to a non-hyphenated married one or adopting an anglicized name, and have experiences and publications that preceded this change, you can show both names at the top using the following formats: Marie Gladstone (neé Smith), PhD, Marie Gladstone (formerly Smith), PhD, or Zhien (Jenny) Chang, PhD. Regardless, in the digital age, you only really need your name, phone, and email. Mobile/cell phone numbers are fine and you can use either a permanent (i.e., Gmail) or institutional email or both. In the US, you should not include any demographic or personal information, photo, or government or school ID numbers. The reason being that this info could potentially bias the hiring process so those involved would rather not have it at all.

Another US-centric convention is having roles, titles, and degrees on the left while the dates are placed on the right. I recommend this for two major reasons: (1) since we read from top to bottom, left to right, the flow of information should follow the same direction. In my opinion, the most important information (roles, titles, and degrees) should be seen first, at the top and on the left. Dates, while important, are not *the* most important details and thus should be relegated to be read last, on the right side of the page. (2) Formatting this way saves *a lot* of space while still allowing for separation of sections. When drawing the attention of the reader, you should use bold, italics, and underline functions as sparingly as possible. I highly recommend bolding your name (especially in publications), titles, roles, and degrees. Within the separate experiences of the CV or résumé, you should avoid large blocks of text and instead use focused, single-line bullet points or references that concisely convey the "who, what, why, where, and how" of your achievement.

When creating a multi-page document, you should use the header and footer functions to include page numbers at the bottom and your name (with degrees) at the top right. Not only does this look very professional, but there are some that still print these out. For instance, if they are not careful, and you have not placed your name and page number on each page, the printouts may get shuffled, and it is your fault (fair or not) that the document is out of order. You should stick to ¾- to 1-inch margins and make sure that your text is consistent in both size and font style. I recommend using 11- or 12-point font with Times New Roman, Arial, or something similar all the while using standard symbols and characters. I know this sounds a bit boring but the more exotic you get with the formatting elements, the more likely you will have formatting errors. Thus, simple elements make for consistent formatting.

Anatomy of the CV/Résumé

The section headings on your CV/résumé are the guideposts for your reader. They should be clearly and specifically labeled as well as centered across the page. While it depends on the specific audience and pursuit, in general, I recommend starting with Education, Research Experience, Awards, Service, Leadership, Presentations, and Publications, in this order, as major sections. In the case of the CV/résumé hybrid or straight résumé, I encourage the following general sequence: Summary, Education, Research Experience, Technical Expertise, Awards, Service, Leadership, Patents, and Publications. Overall, these sections bring the reader to a better understanding of your work, experience, and context.

Typical Section Headings and Their Purpose

Summary = brief overview
Education = qualification
Research Experience = context of work
Technical Expertise = bench/informatics skills
Awards, Service, Leadership = scientific citizen
Patents & Publications = productivity

The "Summary" section is a transitional statement that connects your current and past experience to a new sector or field. This section should only be used on the résumé or CV/résumé hybrid and gives a brief overview of your experience with a few bullets covering the key skills relevant to the job. If you could summarize yourself in one sentence this is where it would go.

The "Education" section outlines only your educational endeavors that resulted in degrees and demonstrates your minimal qualification for the job, i.e., the Bachelors, Masters, and Doctoral degrees. While important, your postdoctoral training is not degree-bearing and should not be listed in this section. It is better placed under your name at the top of the document and detailed in the "Research Experience" section.

Regardless of the type of document, I highly recommend having a "Research Experience" section because it allows the reader to understand the context of your work through descriptive bullet points. This section is divided up based on your current and former research appointments, namely, Postdoctoral Research Fellow, Graduate Research Fellow, Masters Research, and Undergraduate Research. Some of you may even have previous clinical or industry research experience that can be placed here or highlighted further in separate and more specific sections covering "Clinical Experience" or "Industry Experience." How you frame your research experience is important. You have to decide whether to mention every project or just a high-level overview of your interests with accomplishments, for example. Constructing bullet points for research experience for a field application specialist position can be very different than bench position.

Another section specific* to the résumé or hybrid is "Technical Expertise" where you describe your wet lab, computational, analytical, and/or programming techniques under distinct methodological themes. (*As mentioned, you can also use this section in your CV when applying for a postdoc. It falls off the CV when applying for faculty positions because they are assuming you know everything needed for running your lab.) You should use the term "technical expertise" so you do not have to qualify the level of experience you have in each technique; you are an expert in everything you list. I also recommend using this section to collect in one place all the methods you tend to repeat across appointments in the "Research Experience" section. This allows you to stay on topic in each section while emphasizing important information for the reader where they will best understand it. Again, you will have to decide whether to list all the techniques you are familiar with or use a focused or abbreviated approach.

The "Awards" section demonstrates that you are recognized as and contribute to being a well-rounded scientific citizen. You can be very creative with the titles of your sections, and these sections, in particular, lend themselves to descriptive, eye-catching headings. Many of you have different numbers and types of awards that you may want to highlight as one or separately. There tend to be as many as four types: "Fellowships," "Grants," "Awards," and "Honors." Depending on what you want the reader to see, how many of each, and how recent, I recommend using "Fellowships & Grants" and "Awards & Honors" sections on the document. Fellowships and grants may be straightforward, anything you applied for that gave you money to do research. However, awards and honors may be less defined. I would suggest putting travel awards, Dean's list or honor roll, poster or talk awards, departmental or institutional recognitions, and other related honors in this section. If you place something in these sections, be sure not to repeat it elsewhere on the document. That goes for information in all sections.

The "Leadership & Service Experience" section is one of the most effective units you can use to demonstrate your leadership, teamwork, service, and volunteer experience but, unfortunately, it is also the least utilized. Depending on how extensive your experience is in this area, you may consider breaking it out into more than one section covering experiences in "Leadership," "Service," "Outreach," "Volunteer," or any combination thereof. You may even go a step further to distinguish the sector in which you served, for example, an "Academic Leadership & Service" section may include any peer review, departmental committee, or scientific society leadership experiences you have. That said, I must note that this section can and should include relevant non-academic or non-research-related activities and experiences such as coaching youth sports, organizing a 5K run, or sitting on a school board.

Your "Presentations," "Patents," and "Publications" sections demonstrate your research productivity as well as your communication and writing skills. While there are divergent thoughts on where to place them and how to format them in your document, I recommend placing them toward the end based on the reader's tendency to look at the first page and then to flip to the back page. In addition, I advise formatting your patents and publications as long form citations with authors and title of the work followed by the identifying information for the patent or journal. As for your not-yet-published manuscripts, there is a very straightforward means to show that they are indeed in the publishing pipeline. You can either list these as full citations at the top of the section or as a sub-heading called "Manuscripts in Preparation." Regardless, you will follow these citations through the pipeline and label them in progression from "In Preparation," "Pre-Print," "Submitted," "Under Review," "In Revision," "Accepted," "ePub Ahead of Print," to "In Press." After your paper finishes this process, it is now an official publication and needs no further qualifiers. You should note that preprints count as a manuscript in preparation and ought to be listed as such.

In the case of presentations, depending on your career stage and the sheer number of your talks or posters, I suggest that you emphasize oral presentations over posters by having separate sections for them. You may also want to only list a selection of each if you have so many that cataloging all of them would detract from the document. Furthermore, you should only select those that you were either the first author (i.e., primary researcher) or presenting author (if you presented as the primary researcher or on behalf of your research group). In this instance, since authorship is assumed, you do not need to list the other contributors in the reference; you only need to share the title of the talk or poster, the conference or meeting, and when it all took place.

Including a section on "Teaching Experience" or "Mentoring Experience" may be necessary if you are applying for academic teaching-focused positions. They can also be useful in non-academic settings, especially when combined ("Teaching & Mentoring Experience") or expanded to include any supervisory experience you have ("Mentoring & Supervisory Experience'"). I would suggest placing the teaching/mentoring section directly after "Research Experience" and before "Awards" or "Technical Expertise."

When appropriate, you can have other sections and information to tailor your document. I recommend adding "Relevant Training" or "Continued Professional Development" for non-degree granting courses, certifications, and workshops such as a mini-MBA, commercializing science, or Cold Spring Harbor methods course, to name a few. I also advise including a "Collaborations" section for those of you that have many team-based projects across multiple labs, institutions, or countries throughout your research career. Whether you were the lead scientist or cog in the machine, listing collaborations separately unclogs your research experience section highlighting past and current experiences in team science. This is especially important for careers in industry.

For international postdocs, you should also include all the languages you speak or can communicate in using the following sequence of qualifiers: Native, Fluent, and Conversational. In most scenarios, this is enough information and you can always further describe if needed. However, if you have taken language examinations, please share the results here. There are many assumptions made and biases brought by the reader, so, while my general recommendation is to be as transparent as possible, for some, the disclosure (or non-disclosure) of Visa, citizenship, or work status can either help or hinder. The final decision is up to your best judgment. If you think it will be advantageous, then disclose your status, but if you think it might hurt your chances, there is no need to share this information just yet. Ultimately the point is moot because through the application process, employers are going to realize your Visa/work status.

Statements of Research and Teaching

The Academic Research Proposal

Since it is of supreme importance, writing the research proposal can be one of the most in-depth exercises in putting together the academic job package. While constructing the intellectual framework of your lab for the next several years, you have to convey a feasible, fundable, and future-focused research program in just three to five pages to a search committee comprising faculty inside and outside the recruiting department.

Many of you, when drafting your research proposal, will struggle with the scope and level of detail while grappling with not writing a grant, avoiding technical jargon, and providing enough context so the audience can properly assess your credentials and the feasibility of your aims. The early versions of your proposals often lack developed aims where your major question, approaches, methodologies, and potential impact are not distinct nor future-focused enough. Another area you all tend to have difficulty with is providing too much background information and experimental detail leading to a very dense and unwieldy structure.

The half-page summary section of your proposal should define a big problem(s) or question(s) in the field that you will be addressing in several different ways, thus hinting at your aims. You will then depict your past and current research in the remaining half of the page, outlining your major accomplishments and building the foundation of your credentials moving forward. You should mention grants that you were awarded while citing your publications described therein. The last two to three pages of the research plan are used in the framing of your aims, a 10,000-foot perspective that includes both the "smaller" problem (the what and why) and several experimental approaches (the how) that particular aim will address, as well as the impact of having completed that aim. In biomedical research, where I am most familiar, you are expected to develop two to four cutting-edge research aims where, in general, the first aim is usually an extension of your current postdoc work, while the subsequent aims are further afield, mid-to-long term, carry more risk, and can be very creative. Unlike a grant, which this is most definitely not, each of your aims must be independent and on such a scale as to potentially employ several lab personnel to achieve them. It is highly recommended to have an accompanying figure for each aim, or at least one figure per page, to help break up the text along with providing a visual representation of the approach, methodology, or expected data output. It should go without saying, but you should not craft this document without examples, input, and the help of others, especially your faculty mentor.

The Statement of Teaching Philosophy

Drafting the teaching statement can be one of the most esoteric exercises in creating your academic job package. While less philosophical and more practical, the challenge is to communicate a tangible evidence-based teaching approach alongside unpacking your teaching experience in just a single page. The teaching statement outlines your approach, in theory and application, to teaching as well as classroom implementation of active learning and instruction. If written well, this document essentially describes what has and will happen in your classroom due to your teaching interventions, covering inclusive student-centered methods, pedagogy, and lessons learned. As stated, your teaching statement is a mix of concrete and abstract examples, often explained in alternating sequence that demonstrate your command of modern higher education and not just vignettes of passive education where you learned (or were taught) in a certain way as a student from past educators. For those of you that may not have extensive in-class teaching experience, I remind you that teaching is teaching, no matter the format, whether you are leading recitation, tutoring a small group, or mentoring someone one-on-one.

Many of you, when drafting your teaching statement, will struggle mightily with aligning your experience with your philosophical approach. You will find it hard to unearth specific teaching moments while translating your familiarity with research and mentoring to what it looks like in practice in the classroom. Modern pedagogical terminology and techniques are probably foreign to you so you tend to share teaching methods that use the inscrutable primary research literature to demonstrate common scientific concepts. Regardless, there are numerous strategies to tackle these struggles, many being approaches shared throughout this chapter. For teaching statements in particular, you can craft a compelling document by unpacking your classroom content through current events, even connecting with what may be happening in the student's home and school environment. You can also develop a theme for your statement while sharing philosophical approaches and concrete examples that support and reiterate your premise. On the whole, I encourage you to reflect on past teaching and learning experiences and take ownership of how you will teach in the future.

9
Interview Preparation

Becoming a Storyteller

Thus far you have already heard about self-assessment, networking, and updating your materials all with the hope of applying and getting interviews for jobs. Some say, and I tend to agree, that the only way to get better at interviewing is to interview. However, I believe you can help yourself along the way since interview preparation can start long before you are invited to interview. This involves pre-work, understanding your skills, and storytelling. For example, here is an email I received on the topic of storytelling:

> Dear Jim,
> I hope this message finds you well. You may remember me. I met with you once to review my résumé and have attended a few of the postdoc career development events. I also ran into you a couple of times on the bike trail!
>
> I want to share with you some good news: after a somewhat long(ish) process of networking and interviewing, I recently accepted a position at Momenta Pharmaceuticals. I started a few weeks ago working as a Scientist in the Translational Research group.
>
> I saw that you will be leading a seminar on "crafting your story for career transition" in the coming week. I can tell you first hand that for me, it was all about finding and refining that story! In the end, my skills were secondary as compared to what my particular "story" was and how it distinguished me. I truly believe that attending the many postdoc sessions helped me craft this story.
>
> I want to take this opportunity to thank you for all the work you do. Know that it does make a difference, even if people just sit and listen, as I have often done. Something always sticks.
>
> If there is anything that I can ever do to help other postdocs transition, please do not hesitate to reach out.
> Best wishes,
> A former postdoc

This message underscores the importance of interview preparation and the power of storytelling.

The strength of storytelling lies deep in our collective history. Humans are neurologically hard-wired to be captivated by stories and we learn from them by default. To tell a compelling story, you only need three components:

a protagonist, some conflict, and a resolution. Another vital element of storytelling is your audience. A good raconteur captures their listener's imagination by weaving tales around a well-known story framework by providing exposition before moving onto escalating action, in the form of conflict and crisis, ever rising toward a climactic resolution. Once achieved, they explain any lessons that the protagonist learned through falling action. The reason that storytelling is so powerful is that you can demonstrate authenticity, reframe reality, evoke sympathy, convey credibility, build community, forge trust, and wield influence with just a few words.

As scientists, you have many opportunities to develop an appreciation of the importance of stories as you progress through your training. In terms of interviewing, the protagonist of your stories is you, or your research, the conflict is any challenge you encountered, and the resolution is what you did to solve the conflict. Whether you are convincing yourself through inner dialogue or influencing others through networking, you are telling stories of growth, inspiration, and new knowledge. As such, interviewers will ask a lot of questions, thereby giving you many opportunities to share your experiences and points of commonalities through stories. Most of you will get nervous when being interviewed, so I recommend using a methodical approach to answering questions called P-A-R that highlights your actions, behaviors, and competencies. Short for Problem-Action-Result, P-A-R provides a foundational structure for answering questions by telling proto-stories that approximate the traditional story arc by systematically describing your experiences.

Crafting P-A-R Stories

Simply stated, you begin a P-A-R story by defining a problem (P) or challenge that you faced. Next, you specify the actions (A) that you undertook to address the problem. Finally, you explain the measurable results (R) and the impact of your actions. In developing the details for a P-A-R response, I recommend using the "rule of three" as a persuasive technique whereby presenting as many as three examples each for the P, A, and R, allows the listener to better retain the information. You all know your research and associated experiences do not take place in a vacuum and that others are involved, but you still have to tell stories that specifically highlight your role in these situations. Using a simple Problem-Action-Result structure is a very good way to remember how to tell stories about relevant experiences in most interview situations.

For example, your interviewer may say, "Describe a time when you faced a conflict or challenge." Or "Tell me about a situation where you had to motivate others," because that is an important part of your next job. Or they might ask you to explain your most impactful paper or greatest research accomplishment. To this last request, you might reply, "Well, in my recent project the problem we address was [insert Problem here]. The experimental

Interview Preparation

and methodological approaches we used to address the said problem were [insert <u>Actions</u> here]. This resulted in [insert <u>Results</u> here] and we ended up publishing the work at [insert high impact/field-specific journal here] with several citations already." By sharing this P-A-R story, you are demonstrating that the work you are doing is not only important, rigorous, and innovative, but it also has an impact on the field. Below is an example of a response that highlights the P-A-R structure, the relative ease of its composition, and the use of the "rule of three."

Sample P-A-R: Why are you the best candidate for this job at our consulting firm?

<u>Problem</u> (P): Define your strengths – 1) Specific experience in CRISPR; 2) Diverse and broad network of scientists and key opinion leaders (KOL); 3) Experience leading teams in a matrixed environment

<u>Actions</u> (A): Share the process of how you developed something new – 1) Built a complementary team of experts; 2) Established new protocol in X system; 3) Delivered on time with a quick turnaround

<u>Results</u> (R): Talk about results and impact beyond publication – 1) Commercialized new IP and created a company; 2) Invited to give a seminar at a meeting where you are now considered among the KOLs; 3) Received seed funding and submitting new grant

P-A-R Matrix

The strategic combination of the P-A-R storytelling with the certainty that the majority of interview questions can be found ahead of time (through friends, colleagues, and the internet), allows you to prepare your responses before ever landing an interview. In fact, I developed a series of career clinics specifically designed to help postdocs do just that. Additionally, a single P-A-R story can cover multiple topics or skills and may be appropriate to use across several different questions. Furthermore, you can have more than one P-A-R story that describes a certain experience or answers a distinct question. You can begin to imagine that you have actually created a matrix of stories to share at an interview that you can strategically deploy for maximum impact (Figure 9.1). For example, you might be asked a question about leadership and teamwork, and using the P-A-R matrix, you might have developed two or three stories you can tell that fit the interviewer's assessment. To further extend this example, the two or three stories that demonstrate leadership and teamwork may also address your communication skills, ability to take initiative, or problem-solving skills. Depending on the question, you can essentially pick and choose which P-A-R story you want to use at the appropriate time that demonstrates the desired skill(s). While it is important

	PAR-A	PAR-B	PAR-C	PAR-D
Problem solving		X		X
Leadership	X		X	
Teamwork	X			X
Communication		X	X	
Drive to achieve	X	X		X

FIGURE 9.1
P-A-R matrix of stories and skills. (Used with permission from Derek Haseltine.)

to prepare and practice interview responses with P-A-R stories, it is utterly vital that you not commit them to memory word-for-word. You want your reply to feel natural and in the moment and not wooden or rehearsed.

P-A-R Practice

When developing your P-A-R stories, I recommend that you practice your responses with a partner or small group, especially non-native speakers. You should practice with a native speaker to ensure your message is not lost in translation and the tone is appropriate. These low-stakes mock interviews are high-reward situations that allow you to get more comfortable with the flow of details and structure as well as polishing the story. You will find that your answers will naturally lead to more questions from the interviewer (mock or real). Some will ask clarifying questions about details while others will have deeper inquiries that address their interest, confusion, or curiosity. Your first few attempts may be difficult to follow as you are exploring terminology, syntax, and pace while trying to craft cohesive answers. As you practice, your stories will likely get more concise and convincing due to the feedback you receive. Eventually, you will feel more at ease and gain confidence. You may then realize that P-A-R merely provides a systematic structure and that you are the one performing all of the tantalizing elements of memorable storytelling.

Sharing Negative Experiences

Not every question you are asked will be about an achievement or a positive result. Interviewers realize, as mentioned in Chapter 4, that negative experiences and failures can lead to growth and learning in most people. It is quite possible that this will be the basis of an actual interview question. When answering such questions, you must be very diplomatic and not speak negatively about yourself or others. In responding, you need to make sure

Interview Preparation 95

that you are taking responsibility for those things that you are accountable for (perhaps you could have done or said or approached something differently). If it ultimately was out of your control, in a neutral-to-positive framework using the P-A-R structure, you can tell what the situation was, your approach to fix that problem (or how you worked around it or came through it), acknowledge that as a result something did or did not happen, and what you learned from the experience. Below is an example of such a situation.

> **Sample P-A-R: Tell me about a time when you failed to make a deadline.**
>
> Problem (P): Define the failure in three parts – 1) Missed a fellowship grant deadline; 2) Unsure of internal processes and deadlines; 3) Had a competing deadline for a publication
> Actions (A): Share the actions that lead to missed deadline – 1) Plan was not SMART AF; 2) Did not prioritize or say no to other things; 3) Did not delegate or ask for help in time
> Results (R): Talk about results and impact of missing the deadline – 1) No fellowship grant; 2) Delayed project; 3) Postponed manuscript
> Lessons Learned: Share what you learned and will do differently in the future – 1) Took project management course; 2) Incorporated realistic time management/prioritization; 3) Familiarized self with administrative process; 4) Built collaboration with shared accountability

Therefore, you end up turning that negative situation into a "lessons learned" or positive learning experience.

Common Struggles and Successful Strategies for Interview Prep

When it comes to interviewing, you are very concerned about having a natural conversation and the inherent preparation required. Similar to your application challenges, you struggle with talking about yourselves and your research without sounding awkward or salesperson-like. A helpful strategy to deal with that concern is for you all to share concrete, detailed, and authentic experiences through storytelling. However, in the moment, you tend to rush right into answering when you should be trying to slow down and think through and structure what you are going to say. You need to prepare stories that demonstrate soft skills while communicating your leadership approach. Unfortunately, you are often made to feel that getting your research done is your only priority. You do not realize the importance of leadership roles early enough in your training and then, when it is time to interview, you have trouble explaining how you worked with a team, or led a project, both of which are essential skills for setting up your own lab or for joining a research group. Additionally, since many of you may have never contemplated the skills valued outside of academic research until now, you have had little exposure to other possibilities and thus are not comfortable with handling interview questions that pertain to competencies involved in interactive team science. When you remain unsure about your ultimate career direction or are frustrated that you have to apply for jobs that are not your first choice, your dissatisfaction can manifest in your storytelling during the interview.

Having only known the academic environment, you need more knowledge of the sector you are transitioning to, especially industry. Not reading up on the organization and being unsure of who will be attending interviews puts you at a huge disadvantage when interviewing. You also tend to get stressed and question your confidence. It is expected that you will be nervous while interviewing but you cannot let your fear of the unknown result in a lack of confidence. Furthermore, many of you feel too intimidated to take advantage of mock interview sessions, consequently, you end up facing a hiring committee without having practiced at all. Many of you are also unsure what questions to ask and to whom. Some of you may be uncertain how to handle your two-body problem (where a spouse or partner likewise needs employment) that can prevent you from moving in your preferred direction. You sometimes get ahead of yourselves and become afraid of having to potentially choose between jobs or even find it difficult to weigh a single job offer. You also need to understand that the interview does not end when you leave but proceeds through the follow-up and continues until the eventual hiring decision.

Many of the challenges outlined above can be mitigated by thoughtful reflection, preparation, and practice. As I wrote with minute detail in this chapter, you need to prepare to share authentic experiences through P-A-R stories. In doing so, you should also take time to reflect on what your interviewers may really be asking, as if some of their questions have a deeper purpose. For instance, many of you are very anxious that you will be asked about your greatest weakness (a question that is actually not commonly asked). Your interviewer does not want to know your deepest darkest flaw. *If* they ask, they want to know whether you are self-aware enough to acknowledge that you still have room to improve on some important but non-critical area or competency. They also would like to hear that not only are you aware of a weakness, but you are actively seeking to develop new and better skills in that area that might progress into a future strength.

Interview Performance

For the most part, interviews follow a familiar process and structure meant to provide you ample opportunity to make an impression on the hiring manager or committee. You must keep in mind that you may be contacted by different personnel with differing levels of scientific or technical understanding of the job, before and after an interview. In these cases, you should be aware of how to frame your responses (e.g., general HR recruiter vs. hiring manager) which may require adjusting your technical pitch and discussion of experience using or avoiding buzz words, jargon, and field-specific terminology. Of course, certain sectors or fields have specific features that differentiate them from one another. For instance, if you are interviewing at a consulting firm, you will have to prepare for a case-based interview where you are given scenarios and expected to perform back-of-the-envelope calculations that lead to suggested solutions in real time. Many of you may already be familiar with the academic faculty or industry expectation of preparing and giving scientific presentations during your interview; however, you may be less aware that you may also have to give a teaching demonstration or chalk talk. Additionally, depending on the sector and job type, you may be required to go through several rounds of interviews across multiple visits. Regardless, the following advice on interview performance is applicable to all interviews, regardless of sector. Your institution's postdoc office or association as well as its career services office should have ample resources for field- or sector-specific interview preparation.

Do Your Homework

Once you have been invited to interview, you will want to make sure you are doing your homework by exploring the company, department, or institutional website. By researching the leadership team while reviewing and understanding the mission and vision of where you are applying, you will be better prepared to talk about how your values and trajectory align with theirs. With the 24-hour news cycle, you need to be checking the news and social media to see if anything, positive or negative, has happened in between the time of applying and your interview. Upon being invited, you want to make sure that you have a list of interviewers so you can prepare by researching them, similar to when you are networking, to begin looking for points of contact, such as, schools attended, lab pedigree, publications, and relationships in common. When meeting with your interviewers, be sure to point those out and leverage them throughout the conversation. In addition to learning about your skills and experiences, what they want to do is verify and examine what you are saying. For example, they may think to themselves, "They went to this school (or worked in this lab). I went to that same school (or knew of that lab). That school (or lab) has a great reputation." Thus, by association, that means your pedigree (or your research) is respectable. They can also go a step further and actually reach out to their contacts to confirm connections.

The Face-to-Face Interview

You have done your homework and background research on whom you will be meeting with. You have also prepared and practiced your P-A-R stories. Now that you finally made it to the face-to-face interview, there is some additional advice you need to hear for a successful experience. First, I recommend bringing copies of your application materials as well as a small notepad. Ahead of time, you should write notes, reminders, or potential questions for your interviewer(s). I have a terrible memory for names as well as a mental block on certain words, so I create a little cheat sheet for those names and words that might come up in conversation throughout the day. If you have something that worries you, write down any relevant details beforehand to help you address it. Additionally, on the interview day, you will be inundated with information that you may want to keep track of. Therefore, this notepad will come in very handy. It is okay and even somewhat impressive if a job candidate jots down a few notes. Beware, however, to not let it become a distraction for you or your interviewer.

In most interview scenarios, regardless of sector, you will be meeting with several people, from principal investigators, to staff, to chairs, to deans, to CEOs. The first prompt, a non-question, you will probably have to respond to is "Tell me a little about yourself." What I recommend, in addition to your many P-A-R stories, is that you prepare and practice your "career" story. In particular, developing an origin story for your academic career and introduction for yourself, enriched with concrete examples is a great starting point. An effective way to structure this response is to talk about the chronology of your experience from the start (or spark) in science as well as your evolving interest and understanding of science through to your academic experience. You cannot just stop there, however, you also need to explain what led you to that moment of the interview. It is a good idea to prepare versions of your career story that span different lengths of time. This allows you to be flexible and varied in your response (you will repeat this career story to just about everyone you meet with) while helping you sound less rehearsed and more spontaneous. A natural, confident delivery will also help you to engage your interviewer in genuine conversation. It will feel less like you are trying to impress them or sell your ideas and more like you are truly interested in their story and the job. If you are able to move the interview from an interrogation to a conversation then you can ensure that there is a flow where you share but also let them talk and process what you are saying. It is during this back and forth (occurring multiple times with multiple interviewers across several hours) that you will be able to deploy your arsenal of prepared, but strategically delivered, P-A-R stories.

On your end, you have to be (or appear to be) interested in not only what they are sharing, but also observing and interpreting their non-verbal cues while making connections to your experiences. Such visual or non-verbal signals include maintaining eye contact, emoting through facial expressions, assessing body language, and mirroring movements. When you share or respond to a question in this interview conversation you are balancing confidence and enthusiasm with humility and sharing credit with others when it is due. Many of you worry but, I promise, you will not sound like an arrogant salesperson. As you learn more about them and the position, ask the interviewer how you can make their job easier, inquire about what they are looking for in a colleague, what their expectations are, or what their vision is. It is encouraged, and even savvy in my opinion, to turn around and ask the interviewer the same or similar questions they ask you. You should also inquire about what you can expect to happen after you are done interviewing, such as the next steps and decision timeline. Many of you think of the interview as a one-way information dump when in reality you are also interviewing them. The interviewer's responses will influence your decision to accept an offer to work for them just like your responses will influence their decision to move forward with your candidacy.

Gratitude and Follow-Up

After the interview, you should follow-up and be thankful for the opportunity. In fact, it is very important to show gratitude throughout the entire process, from the initial application to the final offer, because once they contact you, you are being interviewed. You should also be polite and show gratitude to everyone who has helped you along the way, including everyone you networked with, who reviewed your materials, and who answered a question. Not only is this good practice and common courtesy, it also makes it much more likely that they will help you again in the future.

Upon completion of the interview process, there is a good chance you will be both exhausted and exhilarated. You may want to follow-up with your interview hosts immediately while the experience is fresh in your mind. I caution you to not do this. Slow down and give yourself, and them, at least a day or two to digest the conversations you had. Take the time to reflect on what you learned, how you felt (then and now), how they responded, and the information you gained. If you send off a quick thank you email right after the interview and you have not had time to reflect, you may look too eager, heedless, or even desperate. When you do follow-up, along with sharing your thanks, you can clarify the timeline and next steps in the interview or hiring process even if that was one of your last questions during the interview. In your "Thank You" email, you need to be specific and very brief. You should say something like, "I appreciate your hospitality and your time ..." while resolving any lingering questions that you have. It is possible, when writing to specific people, to clarify an earlier response or share a link to a paper that you referenced in the interview. In this note, you also want to reinforce your interest in the position (if true) and you can add a sentence like, "Talking with you (or with the team) only increased my interest in the job and I believe because of this conversation and the team that you are building, I would be a fantastic fit for this position."

Unfortunately, there may be a time when they do not respond to or follow-up with you afterward. It is appropriate to follow-up but make sure you give the time to finish an established timeline or fulfill their obligations to any other candidates. As their timeline for follow-up nears, wait until the day after then reach out at that time. "I am just following up to see where you are in the process. I hope it is going well and am hoping to get an update." If you do not hear back, give them a few more days, maybe a week, and then follow-up again. It is fine to reach out multiple times but remain patient because sometimes the process is affected unexpectedly. The only time where it is imperative to push is if you have multiple parallel interviews or offers and you are still interested in that particular position.

In addition to following up with your interview hosts, you may also want to reach out to your references as well as individuals in your network that assisted you by setting up an informational interview. Touch base with them and say, "I just had an interview with [insert employer here], thank you very much for looking at my materials. I appreciate you being a reference, they may reach out to you soon. So you can be better prepared, here are some things that might come up in the conversation."

Finally, I recommend that you reflect on the successes and challenges of this entire job search process, especially aspects of the interview process. Think about what you found easy, tough, challenging, surprising, enjoyable, boring, and so on. Also, consider whether there were certain resources or certain people that you really appreciated (or should avoid next time). Doing this will help you process what just happened but will also help you prepare for the next round of interviews as well as dealing with the offer.

10
Negotiating Your Exit

Considering and Negotiating the Offer

No matter how many applications you submitted or how many interviews you were invited to, you only need one job offer to move on from your postdoc. Once received, you will then embark on a series of new challenges, though most of these will occur with a feeling of optimistic anticipation. Regardless, as with most of your career and job transition journey, you are likely uncertain of the best practices and possible pitfalls of negotiating and accepting a job offer.

While you may not feel like it, you have a large amount of influence in this situation. When an offer for employment is tendered, the employer has given you enormous power and you have the most leverage to negotiate than at any other time in your tenure. Do not abuse this opportunity but negotiate as your best self by assuming good intentions and behaving professionally. You can also dictate the time and place to discuss the offer at length. As you prepare, find ways to relax and increase your confidence beforehand so you are negotiating in the right frame of mind. You should use neutral language, clarify concepts, and take responsibility for your role while staying focused on your goals. Realize that the negotiation process may take place over several days or weeks (for most sectors) and possibly even months (for some sectors such as academia or government). As this is happening, check in with trusted colleagues and mentors to assess your progress, changing tack or altering items to negotiate. Throughout the process, there should be a transparent and ongoing record of the conversation to keep track of what has been decided thus far.

As you prepare for and progress through the negotiation, there are certain principles to which you should be adhering. One thing I emphasize to postdocs is that you do not have an offer until it is on paper (in most cases nowadays this will be emailed as a PDF attachment). You must insist on getting everything in writing as handshake agreements or verbal offers are not binding. Since they expect some level of negotiation, you have permission to inquire about what is negotiable or open for discussion. While you should have clear objectives, you must also be open to give and take as well as being willing to revisit matters for a later time. You should aim to negotiate as a team since they want you to join them as much as you do, but be careful to not over-negotiate as it could harm this nascent relationship and sour them on your candidacy. I truly believe that open, honest, and transparent negotiation is the best policy, however, full disclosure may put you at a disadvantage at times. When this is the case, you do not need to share all your motivations but do not lie, ever. Nevertheless, this process may be arduous and you may need to ask for time to think about and consider your options. In entering into any negotiation, you should know your "walk-away" point, that moment when you have reached an impasse and will not accept the offer unless you get what you are asking for. Remember, you (and they) can say no at any point in the process. It is possible that you might have to turn down an offer (or they rescind one), not for lack of genuine effort, but due to competing priorities or unforeseen changes in circumstances.

In negotiating the conditions of your job offer, you may mistakenly believe that the only thing worth negotiating is salary. While indeed significant, you may be missing out on other, possibly more important, aspects of the bargain by fixating on salary. Besides, your compensation package may not be very flexible owing to budget constraints, so instead concentrate on bettering your quality of life and growth opportunities. It is not about what you are owed but what you are worth. In general, start time, salary, moving expenses, housing allowance, significant other job placement, signing bonus, rate of promotion, management opportunities, expedited evaluation, and professional development are all negotiable. However, many perks such as medical and dental insurance, retirement, and tuition remission are usually fixed.

When you do actually start negotiating salary, try your best to not name a number first. If you have to, name a range that includes your target number (10–20% above and below), based on objective criteria whenever possible. Websites like Glassdoor and public universities list salaries either through crowdsourcing or actual data. Keep in mind that these numbers may be biased or out of date. Salaries vary widely between, across, and within sectors, due to demand, geography, and cost of living among other reasons. Broadly, you can expect a starting academic salary to range from $60,000 to $150,000; $120,000 to $180,000 in industry; over $150,000 in business consulting; and $60,000 to well over $100,000 in science-adjacent or support positions.

The Academic Faculty Offer

As with everything regarding pursuing an academic faculty position, negotiating a faculty offer has its own unique considerations. You will likely be having this discussion with the department chair. In addition to the items listed above, as part of the academic offer, your start date, teaching load, protected time, tenure clock onset, evaluation terms, and service requirements such as committees and administrative duties are negotiable.

If you are planning to run a research laboratory, the most important aspect of the academic offer is the financial start-up package. This is a competitive multi-year commitment from the school to support you and your research that is hopefully enough to cover you and your lab expenses until you are awarded a major grant or cultivate independent funding streams. As part of your preparation, you are expected to know everything you will need to be successful. In negotiating the start-up, you have to make sure that you understand what it costs to manage a lab including the physical space, the number of work benches, cold room access, and culture hoods. The department chair can outline shared resources, common-use equipment, and core facilities available while you can generate a detailed supply budget particularly listing any special equipment. You will also be expected to plan for personnel such as a technician, postdoc, grad student, and yourself as well as how to pay for them.

Regardless of where your next offer comes from, after the negotiation is finished, reflect on the information you gained while clarifying any lingering questions as well as how much time you have to make your final decision. You should be brief and specific in your appreciation, making sure to acknowledge their generosity and time. You ought to also touch base with your faculty mentor, references, and trusted colleagues to seek out their opinions and perspectives. Finally, you must give them an answer after considering whether the negotiated offer will help you be successful in your new job while setting you up for future advancement.

Continuing "The Talk" with Your Mentor

Once you accept an offer for your next position, you still need to deal with and determine many more things. Similar to accepting your postdoc years ago, you again have two bosses. You will need to negotiate a start date (and many other things) with your new manager or chair while negotiating your exit with your current postdoc mentor. Among other things, you will need to discuss how best to wrap up and hand off your project(s) as well as attending to the details of your off-boarding requirements such as data use and materials transfer agreements along with turning over your lab notebooks, handing in your keys, and closing down email access. Perhaps most importantly, this postdoc exit conversation must include the next steps for ensuring that any outstanding manuscripts are finalized and submitted with an agreed-upon timeline and authorship order.

I have outlined an ideal exit situation where you and your faculty mentor are in lockstep with your transition. I highly recommend engaging them early and often in this process. However, some of you may not have a fully supportive mentor or environment or both. In this case, your mentor may or may not know of your job search, eventual job offer, or impending exit from their lab. Whatever your reasoning for withholding this information, you will need to disclose to them that you received, or have accepted, a job offer elsewhere along with a proposed plan for your departure. In order to avoid blindsiding your mentor with the news, you should strive to extend every professional courtesy in giving them notice, regardless of how contentious your relationship may be. You need to notify them and the department at least two to four weeks before your end date so as to expedite the administrative separation process.

Timing Your Exit

Deciding when to begin your next job can be difficult and is influenced by several factors, some of which, like publications, are discussed above. However, the choice could be taken out of your hands altogether if your new employer's timeline and budgets are dictated by the fiscal, academic, or calendar year. Nevertheless, there are further elements you may also have to take into account. For example, you may need to leave your current lab as soon as possible or you have to honor the terms of your fellowship. You might have a partner that is still awaiting their job search outcome or you want your child to finish the school year. You could still have a few months left on your lease and cannot afford to break it or housing is not yet available where you are moving.

Wrapping Up

In making a successful and transparent exit plan with your lab mates and faculty mentor, you will need to prioritize the projects you still need to finish, the pieces of the project you may be taking with you, and the parts of the projects that will continue without you, thus necessitating coordination in handing it off to others. Additionally, the start date of your new job may preclude you from seeing any papers in the pipeline you have (co-)authored to the finish line. If you plan to stay in academia, you may be able to negotiate with your new department chair time to submit, revise, and publish your postdoc papers so that they may count toward grants and possibly tenure. If you are moving to an industry position or any other sector, you might have to do the final steps on your own time or hope that your mentor and lab mates will complete the process without you. I can say with experience that it is very difficult, if not impossible, to publish a paper when you are no longer in the lab. Therefore, my advice is to either finish everything before you leave or reconcile with the fact that this particular part of your work may never be published.

There are many reasons to be on good terms during your transition and to remain so after your exit. You will continue to need references for future jobs and promotions as well as having allies in grant and peer review. The advantage of not burning bridges (even with a rough exit) is that you can maintain, and perhaps enhance, your professional reputation while also leaving open the prospect of collaborating as colleagues later on. It is this collegiality and sense of community born from the evolution of the mentor–mentee relationship to one of relative equals that drive the scientific enterprise more than anything else.

Index

A

Academic cover letter, 78
Academic faculty offer, 102–103
Academia/industry career tracks, 33–34
Accountable goal, 31
Accurate/inspiring story, 44
Achievable goal, 30
Action, 92–93, 95
Activating your network, 67
Advice on application materials, 75
    applications and job postings (*see* Applications and job postings)
    cover letters, 77
        crafting, 78–79
        struggles and strategy, 77–79
    crafting the résumé/CV, 79
        anatomy of, 84–88
        how CV/résumé are read, 82
        organizational and formatting conventions, 83–84
        pedigree, padding and publications, 80
        purpose impacts style, 80–82
        understanding your, 79
    statements of research and teaching, 88
        academic research proposal, 88–89
        teaching philosophy, 89
Analysis paralysis, 37–38
Anatomy of CV/résumé, 84–88
    awards section, 86
    collaborations section, 87
    education section, 85
    leadership and service experience section, 86
    presentations, patents and publications sections, 86–87
    research experience section, 85
    teaching experience/mentoring experience section, 87
    technical expertise section, 85–86

Applications and job postings, 75
    job postings, 76
    stretch positions and multiple applications, 76
    struggles and strategy, 75–76
Authentic/authenticity, 12, 39, 44, 54, 71, 75, 76, 79, 92, 95, 96

B

Becoming a storyteller, 91–92
    crafting P-A-R stories, 92–93
    P-A-R matrix, 93–94
    P-A-R practice, 94
    sharing negative experiences, 94–95
Building new skills, 53–54
Building your reputation, 57–59
Build transferable skills, 59–60

C

Cancer, 7–8
Career, 20, 98
Career transition, 61, 75
Career transition readiness, 65
    activating your network, 67
    common ground, 67–68
    contact points, 68–69
    integrating networking, research, and life, 70–71
    knowing when to leave, 65
    launching your job search, 72–74
    navigating an unknown process, 65–67
    "the talk" with your mentor, 71–72
Changing trajectory via data points, 27
Clueless to clarity, 3–10
Common ground, 67–69
Common struggles and successful strategy, interview prep, 95–96
Communicating productivity, 24

Considering and negotiating offer, 101–102
    academic faculty offer, 102–103
Contact points, 68–69
Convergence of skills, interests and values, 28–29
Cost of living, 22–23
Cover letters, 77
    crafting the, 78–79
    struggles and strategy, 77–79
COVID pandemic, 69–70
Crafting P-A-R stories, 92–93
Crafting the résumé/CV, 79
    anatomy of, 84–88
    how CV/résumé are read, 82
    organizational and formatting conventions, 83–84
    pedigree, padding and publications, 80
    purpose impacts style, 80–82
    understanding your, 79
Create a postdoc trajectory, 26–27
Creating a productive environment, 23–24
Current/future postdoc fellow, 12
CV/résumé hybrid document, 80–82

D

Defining success, 25–26
Developing plans and accomplishing goals
    contemplating the steps, 29–30
    planning and goal setting, 30
    SMART AF goals, 31–32
Developing your network, 67–68
Do your homework, 97

E

Employee Assistance Program (EAP), 39
Environment training, 23–24
Envisioning the endpoint
    changing trajectory via data points, 27
    create a postdoc trajectory, 26–27
    defining success, 25–26
Establishing ground rules and expectations, 47
    meeting expectations, 48
    their expectations, 48
    your expectations, 47–48
Executive level administration, 34

F

The face-to-face interview, 98–99
Faculty mentor, 13, 18, 20, 23–24, 49–50
Fear of "throwing away" your PhD, 42
Finding strength through struggle, 3–10
Following up, speaker via email, 70
Fortitudinous steps goal, 31
Framing your training, 55–56

G

Gaining independence, 26
Getting a job, multiple career paths, 33–34
Gratitude and follow-up, 99–100

H

Harvard Medical School (HMS), 9–10
Hiring process, 19, 84
Human Resources and Employee Development and Wellness Office, 39
Hybrid document, 80–82

I

Ideal job search process, 73–74
Imposter syndrome, 43–44
Integrating networking, research and life, 70–71
International awareness, 23
International office, 23, 39
Interview performance, 96–97
Interview preparation, 91
    becoming a storyteller, 91–92
        crafting P-A-R stories, 92–93
        P-A-R matrix, 93–94
        P-A-R practice, 94
        sharing negative experiences, 94–95
    do your homework, 97

Index

face-to-face interview, 98–99
performance, 96–97
struggles and successful strategies, 95–96

J

Job postings, 76

K

Knowing when to leave, 65

L

Launching your job search, 72–74
Lessons learned, 95

M

Maintaining progress, 32–33
Managing your inner dialogue, 43–44
Manuscripts in preparation, 87
Measurable goal, 30
Meeting expectations, 48
Mentor, 71–72, 87
Mentoring up and self-advocacy, 51–52
Mentorship and individual development plans (IDPs), 49–51
Molecular biology/biotechnology, 3–4
Multifaceted network, 69
Multiple career paths, 33–34

N

Navigating an unknown process, 65–67
Navigating through your postdoc
 establishing ground rules and expectations, 47
 meeting expectations, 48
 their expectations, 48
 your expectations, 47–48
 mentoring up and self-advocacy, 51–52
 mentorship and individual development plans (IDPs), 49–51

professional development, 52
 build new skills, 53–54
 postdoc skills and competencies, 52–53
Negative self-imagery, 44
Negotiating your exit
 considering and negotiating offer, 101–102
 academic faculty offer, 102–103
 "the talk" with your mentor, 103–104
 timing your exit, 104
 wrapping up, 104–105
Network/networking, 6–7, 37, 58, 65, 67–71
Normalizing struggle and failure, 42–43

O

Ombuds office, 39
Optimistic future-focused self-advocacy, 44

P

Padding, 80
Pain points, 37–38
P-A-R matrix, 93–94
P-A-R practice, 94
Path to independence, 55
 building your reputation, 57–59
 build transferable skills, 59–60
 framing your training, 55–56
 preparing for the next step, 60
 structuring your preparation, 61–62
 understanding your priorities, 55
 your training efforts, 56–57
PhDs and postdocs career options, 34
Postdoc job search, 19–20
 interview, 20–21
 location, 22–23
 mentor, 20
 PhD transition, 22
Postdoc needs and worries, 40
Postdoc office/association, 39
Postdoc pain/pivot points, 37–38
 needs and worries, 40
 tough environment, 38–40
Postdoc position, 17–18, 20

Postdoc process, 25; *see also individual entries*
Postdoc protocol, 11, 25
   developing plans and accomplishing goals
      contemplating the steps, 29–30
      planning and goal setting, 30
      SMART AF goals, 31–32
   envisioning the endpoint
      changing trajectory via data points, 27
      create a postdoc trajectory, 26–27
      defining success, 25–26
   getting a job
      multiple career paths, 33–34
   maintaining progress, 32–33
   self-assessment and reflection, 27
      assessing your situation, 28–29
Postdoc skills and competencies, 52–53
Postdoc skills outlook, 18
Potential job search sequence, 73
Preparing for the next step, 60
Problem, 92–93, 95
Problem, actions and results (P-A-R), 92
Professional decision
   creating a productive environment, 23–24
   overview, 17
   postdoc job search, 19–20
      interview, 20–21
      location, 22–23
      mentor, 20
      PhD transition, 22
   postdoc/not postdoc, 17–18
Professional development, 52
   build new skills, 53–54
   postdoc skills and competencies, 52–53

R

Reach/skillsets, 76
Reality and impression trajectory, 26
Red/green flags, 21
Relevant goal, 30
Research and professional integrity office, 40
Research enterprise improvement, 15
Resilience and mental wellness, 44–46
Result, 92–93, 95
Rule of three, 92–93

S

Self-assessment and reflection, 27
   assessing your situation, 28–29
Self-awareness, 28, 29, 43, 44, 57, 75
Sharing negative experiences, 94–95
Situational awareness, 37
   imposter syndrome, 43–44
   postdoc pain/pivot points, 37–38
      needs and worries, 40
      tough environment, 38–40
   resilience and mental wellness, 44–46
   stopping the negativity spiral, 41–42
      fear of "throwing away" your PhD, 42
      normalizing struggle and failure, 42–43
SMART AF goals, 31–32
Some advice on application materials, 75
Specific goal, 30
Statements of research and teaching, 88
   academic research proposal, 88–89
   statement of teaching philosophy, 89
Stealth network, 71
Stopping the negativity spiral, 41–42
   fear of "throwing away" your PhD, 42
   normalizing struggle and failure, 42–43
Stretch positions and multiple applications, 76
Structuring your preparation, 61–62
Struggles and strategy of application, 75–76
Successful postdoc, 25–26
Support systems, 71–72

T

"The talk" with your mentor, 71–72, 103–104
   timing your exit, 104
   wrapping up, 104–105
Their expectations, 48
Time-bound goal, 30
Timing your exit, 104
Title IX and gender equity office, 40
Tough postdoc environment, 38–40

Training, 19–20, 66
Training and career offices, 14

U

Underrepresented groups (URGs), 38
Understanding your priorities, 55
US-centric convention, 84

W

Wrapping up, 104–105

Y

Your expectations, 47–48
Your training efforts, 56–57

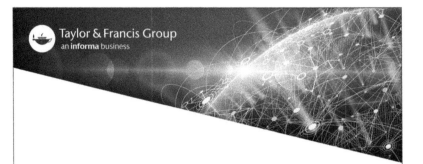

For Product Safety Concerns and Information please contact our EU
representative GPSR@taylorandfrancis.com
Taylor & Francis Verlag GmbH, Kaufingerstraße 24, 80331 München, Germany

www.ingramcontent.com/pod-product-compliance
Ingram Content Group UK Ltd.
Pitfield, Milton Keynes, MK11 3LW, UK
UKHW021056080625
459435UK00003B/27